Missing links in aging

Amalia Gabriela Diaconeasa

Missing Links in Aging

Amalia Gabriela Diaconeasa

Published by Ninsun, 2026.

MISSING LINKS IN AGING

First edition. February 26, 2026.

ISBN: 979-8233921995

Written by Amalia Gabriela Diaconeasa.

Table of Contents

INTRODUCTION

Aging is the greatest problem of the human kind, as some of the oldest literary works show it, depicting the feelings of despair and horror people experience when they face it, as well as the pursuit of eternal life and youth as the only hope of defeating it. In the Ancient times, a dangerous and violent period of our history, death could have a lot of causes, but the ultimate cause of it was aging, which gave no chance to any human being.

Then, most of the time, the solution was seen as residing in magic and the supernatural. But now, it comes to modern science, the real and powerful wonder weapon, is the only one which can provide a viable solution. Maybe solving aging is the most important scientific challenge, although most medical researchers do not think it is possible, and aging is not even considered a disease, but a natural process embedded in our genes. Some researchers even propose a putative genetic program of aging. Any treatment must be more or less palliative, not addressing the ultimate roots of aging.

Unfortunately, we are not traversing a period of challenging nature, but submitting to her mysterious laws. In order to submit to nature's laws, we have to know them.

Aging has intrigued me since childhood. I was in 3th grade when I decided what I wanted to do in my life: to solve aging. Being born in a country where the first institute of gerontology and geriatrics, as well as the first anti-aging product was created, encouraged me and made me think this endeavor was worth pursuing. Beyond any therapeutic and scientific value, this product and this institute inspired me believe that a cure for aging was possible, but first of all, a scientific approach of aging was very close to me. Ana Aslan, who founded that institute around an empirical discovery of some therapeutic effects in age-related diseases of a local painkiller (procaine, chemically related to cocaine), would have done so much for gerontology, even if her only success was to transform aging into a medical problem (and business).

I studied biochemistry in order to become researcher and study aging. Biochemistry was ideal, because it is about life's deepest mechanisms, I thought. After I graduated and got a master degree in neuroscience, I got a job at the "Ana Aslan" institute and started to follow my dream. Since then, every day of my life I have not stopped thinking of aging and its mechanisms, trying to understand it.

But in order to understand aging, one has to understand life. And the backbone of life is evolution. Life is, first of all, evolution.

Starting from the study of aging, knowing the effects of caloric restriction, I came to the idea that this diet could have been an evolutionary force, at least during human evolution. I developed a new hypothesis of human evolution, based on a complex analysis of the physiological, genetic, biochemical differences between *Homo sapiens* and the most related species, *Pan troglodytes* and *Pan paniscus*. In 2007 I published a book that presents this hypothesis, The Civilization of Famine/The Human Evolution Seen from Another Angle.

Homo sapiens suffers from age-related or degenerative diseases that are rare or absent in the most related species (*Pan troglodytes* and *Pan paniscus),*which are type II diabetes melitus, cancer, autoimmune disorders, psychiatric disorders, autism, Alzheimer's disease. A hypothesis able to explain human evolution should also be able to explain the incidence and development of these disorders in humans.

One of the most important biochemical pathways altered by caloric restriction, (which also can be involved in human evolution, including the appearance of some specific human traits), IGF1 (insulin-like growth factor 1)-insulin pathway, is also altered in these disorders.

Human evolution is just a species' evolution. But the principles of evolutionary changes, in biochemical terms, if verifiable, could be applied to any species. Aging is an evolutionary process, obvious in some species, but slow or negligible in others.

Based on these ideas, in 2015 I published a new hypothesis of aging, on evolutionary bases. although I developed it in 2012. This book is reflects my vision on aging. It is based on some articles I published during my PhD study. but not only. It starts with the intriguing facts related to aging I noticed in my lab, the developing of the new hypothesis of aging, which I called the biochemical hypothesis of aging, continuing with its theoretical predictions and implications related to other aspects of aging. This hypothesis was also experimentally exploited. New treatments targeting longevity, but also some degenerative disorders, were tested. A longevity study in mice had promising results. Some substances able to change the pace of aging in mice, could have anti-aging effects in humans and a patent based on these results could be filed. Another experiment I conducted suggests that the fertility in old mice could be restored.

But this book is only the beginning of a process of developing new experiments, new theoretical implications and maybe new applications. Aging, beyond its scientific aspects, has huge social and economic implications. A solution for prolonging life and delaying aging is mandatory, taking into account the economic burden of the aging population. Although most politicians are not aware of this problem, a cure of aging is the best solution for this problem.

A NEW HYPOTHESIS OF AGING

Introduction

Since ancient times, as recorded in the Epic of Gilgamesh, people have been trying to find a solution to counteract the aging process. Here we will offer a critical analysis of some important data and hypotheses related to aging, as well as present our own hypothesis.

What is aging? How can we prevent it, and can we reverse it? There are a lot of hypotheses of aging (e.g., the free radical hypothesis, the hypothesis of neuroendocrine phenomena, the collagen hypothesis, and others), but none of them has a great predictive power. The experimental data available are often contradictory when it comes to supporting one or another of these hypotheses. The best known is the free radical hypothesis, and it is also the hypothesis on the basis of which most experiments were conducted and the largest amount of data was collected. It is a 'gift' of radioactivity, in the sense that it was radioactivity that inspired scientists to come up with this hypothesis. After the harmful effects of free radicals on health became known, it was presumed that free radicals, which appear in the course of some cellular reactions (such as mitochondrial respiration), but whose production is intensified in conditions such as inflammation or even obesity, might be responsible for the phenomenon of aging as well. But even this hypothesis has contradictory predictions. If free radicals are responsible for aging, then subjecting an organism to massive irradiation should result in accelerated aging. Indeed, radiation does affect cell division and the effective functioning of some tissues and organs, and it is responsible for grave illnesses and mass genetic mutations.

Exposure to radiation for therapeutic purposes leads to the acceleration of some processes that can be blamed on aging. For example, photoaging (the action of UV rays on skin with a low amount of melanin) is responsible for fast skin aging in Caucasian people. Melanin has an antioxidant effect, it protects against UV rays by the nature of its structure. Exposure to radiation, for instance used in therapy, accelerates some aging processes in mammals such as humans, but the freshwater hydra (an organism composed of stem cells), when irradiated to the total destruction of its capacity for cell division, still regenerates, as from an irradiated and severed hydra results a smaller hydra, through cell redifferentiation [1].The hydra has a type of regeneration that is different from that of vertebrates, but this experiment shows that aging does not depend only on an organism's capacity for cell division, but also on other biochemical processes.

In addition, although there are data that suggest that some antioxidant systems (antioxidant enzymes like catalase, superoxide dismutase, and glutathione peroxidase) are affected in the aging process, which would allow free radicals to attack, there is no correlation between the lifespan of a species and the efficiency of its antioxidant systems. Birds, a class of animals with high life expectancy, do not have powerful antioxidant systems, in all probability precisely because these organisms do not produce large quantities of free radicals, for instance as a result of mitochondrial activity [2]. Even more interesting is the fact that this situation holds for mammals that live for a long time, taking into account body weight. Naked mole rats (*Heterocephalus glaber)*, rodents the size of a mouse which live for up to 20 years, have less efficient antioxidant systems than other rodents with a lower life expectancy [3]. Given these data, one may ask if, even considering the idea that the production and action of free radicals rises as an organism ages, their level can be so elevated as to produce the known phenotypic effects. Recent data weaken more and more the connection between free radicals and aging. In *Caenorhabditis*, the mitochondrial mutations connected to a reduced capacity to counteract the action of free radicals are actually associated with a longer lifespan [4]. Moreover, Paraquat, a toxic herbicide with a high capacity to produce oxidative stress, increases the life duration of these worms, whereas antioxidants appear to curb these effects [4]. The superoxide overproduction present in these mutants contributes to the increase in their life duration, through a different path than others involved in longevity. Aside from being the main source of free radicals, the mitochondrion is the source of cellular energy. Numerous data suggest that affecting the mitochondrial function leads to accelerated aging. Recent findings show that some old and obsolete drugs used for treating AIDS especially in third world countries produce accelerated aging by affecting the mitochondria [5]. Mitochondrial mutations accumulate on a large scale, but it seems this is just a faster turn-over of the mitochondria, which through poly-cloning, leads to a proliferation of already existing mutations. Still, stimulating mitochondrial activity in *Drosophila* through overexpression of the homologue of the mammal transcription cofactor PGC-1 (dPGC-1) in intestinal stem cells leads to an increase in life duration and slows the effects of aging [6]. A mitochondrial mutation associated with the alteration of the functions of this organelle in *Caenorhabditis,* Mit, produces an increase in life duration [7]. This mutation leads to a special metabolic footprint, distinct from aerobic growth. When the respiratory chain is affected, alternative energy sources are sought.

What is more, when energy production is affected by removing an enzyme that is important for energy metabolism (fumarate reductase), a new increase in life duration is observed. All these contradictory data on the effects of free radicals in aging and on altering the mitochondrion's function as a cell power plant are interpreted in what follows in light of the new hypothesis of aging.

What modifies the aging rhythm?

In order to build an image of aging as such, we will discuss some of the most important experiments on this process. The puzzle of aging could be solved if we take into account the conditions that favor accelerated aging, slow aging, or the disappearance of this phenomenon.

Rapid aging syndromes, under the generic term progeria, are genetic conditions given by mutations that often affect the integrity of genetic material, codifying enzymes involved in the architecture and repair of said genetic material [8]. Still, the genetic defect in the most spectacular Hutchinson–Gilford progeria, in which the aging rate is 8–10 times bigger than normal, is a protein in the nuclear membrane, lamin A [9]. A point mutation of the gene responsible for codifying this protein leads to a deficiency in its processing and its accumulation in the nuclear membrane, which takes on the appearance of a blebbing. The connection between these blebs and the modifications similar to aging they generate is not currently known. What is certain is that this protein does not appear to have a significant survival role, as KO mice for this protein exhibit normal embryonic development [10]. Also, it seems that lamin A accumulates in normal aging as well, even if this accumulation has a slower rhythm than in progeria. What is interesting is that some studies suggest that lamin A may be responsible for genetic modifications associated with aging, such as DNA defects, defects of the methylation of histones, which influence heterochromatine, and others [11]. The mutant version of lamin A also appears in cells from young, healthy patients, after several passages, and its accumulation in the nucleus in older healthy patients that is not accompanied by a rise in the mRNA responsible for its synthesis suggests that the protein processing apparatus of the cell might be to blame for this accumulation. In other words, we may speculate that the abnormal functioning of other proteins may lead to this phenotype in the case of aging.

The way in which the aging process is slowed in a wide series of phylogenetically distant organisms (from yeast to worms and even mammals) is caloric restriction (CR). CR represents a diet with all the necessary nutrients, but with a reduced caloric intake (for example, 60% for rats). The main target of CR is the signaling pathway insulin/IGF1 (insulin-like growth factor 1). Blocking a receptor equivalent to the mammalian insulin receptor boosts the life duration for *Caenorhabditis* worms by 50%. Recent data show that CR also acts on a nuclear receptor, NHR-62, with implications on lipid metabolism, which is modified in conditions of CR [12]. CR would also be involved in autophagy, which is likewise important in life prolongation. Until recently, CR was considered to be the only way to prolong maximum and average life expectancy for a large number of phylogenetically distant organisms. However, more recent experiments show that there also exist substances that could lead to similar results. Rapamycin, an immunosuppressant (actually an antibiotic), given to 600 day old mice – an age at which mortality touches 90% – lead to an increase in survival rate of 13% for females and 9% for males [13].

A different study suggests that the effects on life duration would also appear in mice that are given this drug at 270 days, when mortality reaches 51% in females and 68% in males. What is interesting is that rapamycin seems to modify the phenotype of progeria-affected patients, reducing those blebs on the nucleus [9].

Its target, m[TOR], is a protein that regulates cell growth, survival, proliferation and motility, protein synthesis and transcription. It is also involved in autophagy [14]. m[TOR] is part of two molecular complexes that act somehow downstream from the insulin/IGF1 pathway, meaning its role would be as a regulator of this pathway [15]. What is interesting is that another molecular complex is connected to MAP kinase, which is involved in the formation of the cytoskeleton (microtubules). The cytoskeleton is altered in numerous degenerative illnesses In Alzheimer's disease, those neurofibrilary tangles are made up of aggregates of hyper-phosphorylated *tau* protein. The *tau* protein intervenes in the formation of the microtubules. It would appear that hormesis (moderate stress) is a factor that influences aging about whose mechanism we know nothing.

But there is stress even at the cellular level. Experiments conducted by us and others [16–18] show that the cytoskeleton is temporarily altered in cellular stress (cold, nutrient deprivation, toxic substances, etc.). Most importantly, actin stress fibers disaggregate, but other fibers are affected as well. We might speculate that the cytoskeleton, from which later resulted contractile structures (from flagella and cilia to muscle fibers) 'feels' the energy state of a cell and changes accordingly, which influences cellular motility and other functions derived from this. It may be deduced that the signal given by the changes of the cytoskeleton, which are similar to those that arise under stress conditions, may lead not only to external modifications (of cell or organism motility), but to internal ones as well, given by the synthesis of some substances or the activation of some genes.

Could we, in this manner, transmit to a cell the information that it is facing problems, and that it has to synthesize certain substances and activate certain genes in order to react to specific environment conditions? Treatment of striated muscle cells with myoseverin, a purine that disintegrates microtubules, shows that it can determine *in vitro* growth of muscle cells after this toxic substance is removed [19]. Another purine, reversine, also has regenerative effects [20]. We suggest, these purines with strong antimitotic effects (myoseverin has been recommended for treating cancer) could determine modifications in gene expression, thus modifying the cellular phenotype in the same way as that of cellular stress. The antimitotics used for treating cancer, with their toxicity reduced through various

methods, might have anti-aging effects. In fact, 5-fluorouracil, a pyrimidine with use in oncology, when it was used topically for treating skin tumors, also had the effect of attenuating wrinkles [21].

Another way to increase life duration, this time in yeast, is manipulating a gene that codifies an enzyme with effects on DNA architecture. This gene, sirtuin, reduces recombinations, and has epigenetic effects [22]. This gene is also present in mammals (SIR2), and it is assumed to have similar effects. At the cell level, the enzyme called telomerase, which replicates the ends of chromosomes (telomeres), had aging-reducing effects without malignization. Malignant cells secrete telomerase, and if they do not, they replicate their telomeres anyway, through recombination. Telomerase has been the subject of much literature, and a Nobel prize was even given for work on it. It is known that telomeres decrease in size with every cell division in the absence of telomerase. Only this enzyme appears in cells that divide quickly, such as those in the mucosa, and aging affects them as well. What is more, there are organisms that do not live for a long time and secrete telomerase constantly, even though telomerase is secreted in small quantities in human fibroblasts, a fact which inspired its connection to aging [23].

Fast or slow growth within the same species may lead to great differences in lifespan. An example is the white cedar, *Thuja occidentalis*, which lives for 80 years in cities, but for up to 5000 years on steep slopes [24]. An interesting aspect about aging is represented by the experiments that show that blood transfusions from young animals to old ones have regenerative effects [25].

The way to erase the effects of aging is meiosis. However old the parents might be, children are invariably born young. What is happening? Experiments on yeast show that bringing them back to the meiosis phase cancels the effects of aging. It would seem that some transcription factors are involved, and, indirectly, gene expression [26]. What is interesting is that meiosis is a process capable of repairing all DNA defects, including the most difficult ones, such as double-strand breaks. There are enzymatic 'keys' for all DNA errors, but for every one of them there are certain conditions in which different reactions of this type are favored more or less. Contrary to what was previously thought, it is not the case that in aging there is an accumulation of a large number of random mutations. Just the opposite, mutations affect certain specific genes, such as those involved in repairing mutations [27].

A model of aging

In order to develop a model of the aging process, we must first develop a summary model of the processes involved in life. According to this new model we have imagined, life is a fabric of reactions that stimulate and perpetuate each other. The interaction of living organisms with the environment is translated into biochemical modifications that may have a certain temporal inertia which varies according to the biochemical environmental factors specific to each organism. In other words, a modification of a biochemical reaction, at the level of codification (epigenetic modifications) and regulation, is maintained for various amounts of time. This model starts from a few simple postulates, based on observation. The simple biological principles would be as follows. Life is a sum of intertwined chemical reactions, which have effects at a distance within an organism and even in the external environment. They are, however, subject to the following phenomena: (1) A limited number of reactions may take place within eukaryotic cells. This limit is given by the energy limitations of the respective cells; (2) the reactions that take place are favored by the existence of reactants (including extracellular signals), and depend on energy support. The differentiation of organisms was accompanied throughout the course of evolution by the limitation of the number of reactions that take place in the eukaryotic cell. Eukaryotes have probably evolved as a consequence of the tendency towards minimum energy of thermodynamic systems. Likewise, pluricellular organisms would have evolved as a consequence of new ecological constraints, leading to new points of minimum energy.

According to which reactions are favored, various cellular states will appear, both normal and pathological (e.g., degenerative illnesses, cancer, etc.). This phenomenon may also intervene in speciation (the appearance of new species).

Given this model, aging is a loss of the functional capacity of an organism, probably of separate tissues. It is not just the loss of the ability for replication of its cells. There are tissues that grow slowly, and their stem cells do not participate in regeneration activities anymore.

According to this new hypothesis, aging is a shift of global reaction equilibria, in the sense that cell differentiation (specialization) is erased. In other words, the transcribed information and the concentration of certain substances synthesized by a cell are literally diluted. This has effects even on the phenotype of distant cells, effects that are especially visible when the cell in question is part of the nervous or endocrine system. Aging is a result of a diminution in the synthesis rate of specific protein substances The substances are produced in smaller quantities. This is in agreement with epigenetic modifications that appear in aging, meaning that expressed gene specificity is lost [22]. In other words, more genes are transcribed, but probably in smaller quantities, and tissues lose their clear differentiation. It is relatively easy to verify this supposition, through simply measuring the levels of certain enzymes or specific proteins or mRNA that codifies them. The data on skin and muscle aging, for instance, are in agreement with this aspect. Likewise, the role of sirtuin in aging is in accord with the model. Sirtuin contributes to maintaining cell differentiation and maintaining the architecture of genetic material, and therefore maintains genetic information.

The substance quantities are reduced as a consequence of reduced cellular stimulation, which leads to a diminution of the signals that the cell transmits at a distance. At the level of the cell itself this process has long-term consequences. The 'energy allotment' is dependent on the reactions that take place. As a comparison, the effects of physical activity on health are well-known, but this is just a facet of a larger phenomenon. If there are no specific reactions, or their level is low, the 'allotment' is reduced, or other reactions receive energy, such as those connected to cell division or cellular support (cancer, degeneration).

On the other hand, a lack of stimulation reduces cellular division, since there will not be reactions or syntheses that support the capacity for cell division. In this global phenomenon, an important role will probably be played by the degeneration of support tissues, which affects highly specialized cells. The growth of an organism would be phylogenetically limited by the decrease in such reactions.

This model is in agreement with the older prediction presented in previous papers that aging is affected not only by the loss of stem cells (to a smaller degree), but also by the differentiation defects that appear in their case. A recent paper shows that in androgynous alopecia there are stem cells, but they do not receive signals that instruct them to differentiate [28]. The environment is very important for the well-functioning of stem cells, as they also receive signals from surrounding differentiated cells [29]. Recent data confirm this prediction through the fact that a growth factor involved in differentiation is responsible for the rejuvenation effects blood from young animals has on old animals [30]. Another aspect that is essential in aging is the existence of energy support. This explains the influence mitochondria have on aging. When there is energy support (high signaling on the insulin/IGF1 pathway, for instance), reactions that produce degeneration (such as in the case of cancer, and not only) may be coupled, meaning that there will be 'energy allotment' for them. In the absence of specific reactions, other reactions will be favored. Reactions that normally do not appear may be coupled energetically, with effects on pathology and not only.

Actually, in conditions of shortage, division reactions are inhibited, and specific and maintenance reactions are favored. This could explain the effects of CR, which may favor energy accumulation

reactions, at least to some degree, as well as energy consumption. Indeed, autophagy is stimulated in CR. Stimulation of reactions involved in the access to cellular energy is essential to cell functioning and specificity.

It is possible that certain types of cellular reactions are 'financed' differently, meaning that they are favored, according to the type of cellular respiration: aerobic or anaerobic. Cell division would be favored by anaerobic respiration, while aerobic respiration would favor specific differentiation reactions. It is not by chance that, during evolution, differentiation appears in eukaryote organisms that have mitochondria (and plastids, in the case of plants). Abnormal division, like the one taking place in the case of cancer, is associated with a disturbance in aerobic respiration.

Mitochondrial activity may be involved first of all in 'financing' reactions that synthesize proteins which have the role of 'relating', meaning that they are more or less destined to the functioning of the cell within its organism, not just its own division and maintenance, reactions 'financed' mainly by glycolysis.

It is no surprise that, in general, animals with negligible senescence grow indefinitely. The growth of organisms is characterized by certain reactions that have a 'regenerative' role. During growth, a greater quantity of substances involved in reparatory processes is synthesized, making growth a kind of extended lesion. A piece of evidence in this regard is brought by experiments on mutant nude mice. When they had deep wounds on their backs, they got their hair back [31]. The WNT gene, which is involved in embryogenesis, could be responsible for this process.

Why do mammals not live for a very long time, as do some reptiles and fish? With mammals, not only are reactions very fast, but they are also more non-specific than in reptiles, for example species that are capable of regeneration (the axolotl, the newt) or exceptional longevity (some species of turtles). When reactions are fast, as they are in the case of homeothermic animals, there is a greater chance that their specificity will be lost. Reversible thermal transformations are quasi static. The 'inertia' of biochemical reactions is greater in some species than in others. Species with a slower metabolism, as well as those with a greater body mass would have higher biochemical reaction inertia.

Predictions and explanations of aging-related phenomena offered by this model

Our hypothesis answers a great number of problems related to aging, as well as answers most observations and experiments. First of all, it explains the unequal aging rhythm observed in some species of mammals, especially humans, as well as the devastating effects of andropause and menopause on health. The hypothesis would explain why there is a direct correlation between maximum life duration and body mass, and also metabolic rate, especially in mammals. It also explains why slow growth is associated with great differences in longevity within the same species, as in the case of the white cedar, which lives for 80 years when it grows in the city, but probably for 5000 years when it grows on rocky mountain slopes [24]. Likewise, the hypothesis is in accordance with the lifespan difference between birds and mammals, with the differences between bird species that fly and those that do not, and also with the longevity of the bat, a flying mammal with a lifespan similar to that of birds [32]. Homeothermic animals that fly, and therefore subject their bodies to constant energetic and functional constraints, despite their superior metabolic rate (birds), sustain certain mandatory reactions. A small mammal that does not fly, like the *H. glabe*r (the naked mole rat), probably lives for as long as it does because of stress (demands) brought by crowding, infections, lack of sustenance, and complexity of social relations. Issues related to the reproduction of this small mammal may also have some significance. A certain heterochrony, the alteration of prenatal development stages (neoteny), might provide the explanation. The lack of hair, certain aspects of its reproduction (the inhibition of reproduction in most individuals, but its exacerbation in the queen) recommend it as a neotenous organism [33].

The strong inhibition on contact of cells in cultures [34] suggests the existence of reactions associated with maintaining cellular structures and their high level. Neoteny would involve maintaining reparatory and other reactions that appear during growth.

The model would, however, also explain why, within the same species, individuals with lower body mass live longer. This is the case for humans as well [35]. Within the same species, differences in metabolic rate are small, but in smaller animals the 'dilution' of reactions would be more reduced. For larger animals of the same species, degenerative illnesses could be higher in frequency, which

seems to hold for humans. A high level of IGF1 (insulin-like growth factor 1) is associated with an elevated risk of developing degenerative diseases [36]. Moreover, the mechanism of progeria, in which cellular signaling (and thus gene transcription) is reduced, seems obvious. In the case of progeria, cells seem to divide faster, which affects their specific syntheses. What is more, associated pathology (heart conditions, degenerative and autoimmune illnesses) would stem from the reduced level of some specific signaling, and the reduced level of some proteins involved in cellular signaling (with modulating effects on immune response, for example). The model also explains the mechanism of caloric restriction, which may be said to 'clear' intracellular reactions, in the sense that it orients reactions towards the maximum efficiency of the organism as a whole, favoring intense feedback reactions. At the same time, it reduces inflammatory processes, which are important in aging, and more so in progeria.

Likewise, the Hayflick limit (the limited division of cells in a culture), which has a certain connection with the longevity of a species (higher in humans than mice, but lower than that of turtles) [37], but which differs according to the tissue of the host, can be explained better than through the effect of telomere shortening. Orienting the metabolism towards reactions that are as slow and long as possible, and that have many implications, stimulating other reactions, can have the effect of preventing aging.

The view on aging as a process of denaturation through functioning (for instance, free radicals resulted from oxidation reactions) could change into viewing aging as denaturation through non-functioning. Some reactions do not take place anymore, or their level is lower.

Aside from these explanations, the model foreshadows ways of combating and preventing. Analyses may be developed that show the level of some metabolites, which then allows certain reactions to be stimulated. This could be a way to diagnose diseases like Alzheimer's, which are currently problematic from this point of view.

Likewise, the 'longevity genes' could turn up in surprising places, in the sense that certain genes involved in metabolism and cellular signaling (not just in repairing DNA defects or in the architecture of the genome) could create significant differences in life expectancy. A molecular form of an enzyme involved in sugar, lipid or protein metabolism, or of a cellular receptor, for instance, might make a difference in longevity. The treatment and/or prevention of aging or degenerative diseases could be done using various methods, which would include simulating cellular stress (for instance with the aforementioned purines) throughout the entire body, and in vitro stimulation of some cells, followed by their reintroduction into the body. Stimulating specific reactions in certain tissues is another such method.

The biochemical hypothesis of aging is in agreement with some particularities of aging in vertebrates and not only. A prediction of this hypothesis is that different parts of the same organism may age following different paces. Older and newer data support this assumption. Telomere length varies across peripheral lymphocytes from the same individual, between similar cells of different individuals of the same age, and across cell cultures of fibroblasts of the same donor [22]. The replicative capacity and telomere length are influenced by stress factors, disease. On the other hand, epigenetic changes related to age have different paces in different tissues [39]. For the most tissues the chronological age matches the organism's biological age (a statistical approach) but for some specific tissues, the DNA methylation of some specific markers correlated with age undergoes much faster. Cancer prone tissues and tissues situated near tumors ages much faster than healthy cancer free tissues or tissues with lower risks of malignization.

Aging in bacteria is also in accord with the predictions of this hypothesis. Until recently, it was considered that bacteria don't senesce. But cellular senescence, even in the presence of abundant nutrients, has now been demonstrated in several microbes, including most recently the bacterium *Escherichia coli*. With respect to microbes, the critical measure of aging would be a reduction in reproductive capacity — longer generation intervals resulting in fewer offspring produced as a function of time. Asymmetric division in microbes, from the eukaryotic budding yeast, *Saccharomyces cerevisiae* (which most work on microbial aging has been used) to *Caulobacter crescentus* (a bacterium that undergoes asymmetric division, unusual in prokaryotes) was associated with aging of the "old" pole [40]. Over the course of roughly 100 generations, the mean generation time nearly doubled, from approximately 2.6 h per generation, to over 5 h per generation. But this situation was true only for the cells of one particular pole. Despite descending from cells whose reproductive output was already in decline, these new cells of the "new" pole divided as rapidly as the cells at the beginning of the experiment. Even a superficially symmetrical division, as in E. coli, produces two distinct cell poles, which "age" differently. The magnitude of the loss of reproductive potential was similar to that observed in *Caulobacter*. Aging in *Escherichia* was refuted by other experiments, suggesting that death of *E. coli* is not purely stochastic but it is the result of accumulating damages [41]. This discrepancy may be due to the different culturing methods used in the studies. Although symmetric division is not associated with aging in microorganisms, the decline in growth rate in *E. coli* [42] appears to be at least partially attributed to the asymmetric distribution of some metabolites, preferential localization of inclusion bodies near the old cell wall [43]. This localization is thought to be the passive result of the slow diffusion of the large aggregate, and the exclusion of the aggregate by the nucleoid [44]. A similar mechanism of aging has been found to occur in *Schizosaccharomyces pombe*, which divides via symmetrical binary fission [45].

Under favorable conditions, few cells died, and their death was random and sudden rather than following a gradual increase in division time. Cell death was correlated with the inheritance of Hsp104-associated protein aggregates, due to exposure to stress factors. After stress, the cells that inherited large aggregates aged, were showing a consecutive increase in division time and an increased death probability. Their sisters, who inherited little or no aggregates, did not age.

These data, which strongly agrees with the biochemical hypothesis of aging, illustrate very well what aging may be, i.e., altered chemical cellular profile resulted from lack of specific reactions, which can be restored in some condition. The new theory of evolution could indicate possible plant, fungus and bacterium species given their ecology, where one could discover substances that have an effect of cellular rejuvenation.

Of course, illnesses associated with aging (from type II diabetes, cancer and cardiovascular illnesses, to rheumatism) would benefit from new treatments. Most of the pathological modifications associated with aging are reversible. Aging could be not only prevented, but also treated. Proof in this sense is the fact that KO mice for telomerase treated with telomerase not only exhibited a slower aging rate, but the premature effects of aging started to become attenuated [46]. Mice are organisms with a high level of telomerase. But telomerase is just part of the issue. A confirmation of this hypothesis comes from recent data that suggests the restoration of mitochondrial oxidative phosphorylation affected by aging with a rise in the cellular level of NAD+, a phenomenon that depends on SIRT-1 [47], after a pseudohypoxic state was responsible for the loss of communication between the mitochondrion and the nucleus because of aging

In conclusion, aging is, in light of this hypothesis, caused by a decline in functioning/stimulation. Given these conditions, the possibility of its reversal is discerned.

Parts of this chapter appeared in Diaconeasa, Amalia Gabriela, Mariana Rachita, and Raluca-Ioana Stefan-van Staden. "A new hypothesis of aging." *Medical Hypotheses* 84.3 (2015): 252-257.

References

[1] Hicklin J, Wolpert L. Positional information and pattern regulation in hydra:

the effect of y-radiation. Embryol Exp Morph 1973;30:741–52.

[2] Floyd RA, West M, Hensley K. Oxidative biochemical markers: clues to

understanding aging in long-lived species. Exp Gerontol 2001;36:619–40.

[3] Andziak B, O'Connor TP, Buffenstein R. Antioxidants do not explain the

disparate longevity between mice and the longest-living rodent, the naked

mole-rat. Mech Ageing Dev 2005;126:1206–12.

[4] Yang W, Hekimi SA. Mitochondrial superoxide signal triggers increased

longevity in Caenorhabditis elegans. PLoS Biol 2010;8:e1000556.

[5] Payne BAI, Wilson IJ, Hateley CA, et al. Mitochondrial aging is accelerated by

anti-retroviral therapy through the clonal expansion of mtDNA mutations. Nat

Genet 2011;43:806–10.

[6] Rera M, Bahadorani S, Cho J, et al. Modulation of longevity and tissue

homeostasis by the drosophila PGC-1 homolog. Cell Metab 2011;14:623–34.

[7] Butler JA, Ventura N, Johnson TE, Rea SL. Long-lived mitochondrial [Mit]

mutants of Caenorhabditis elegans utilize a novel metabolism. FASEB J.

2010;24:4977–88.

[8] Best BP. Nuclear DNA damage as a direct cause of aging. Rejuvenation Res.

2009;12:199–208.

[9] Cao K, Graziotto JJ, Blair CD, et al. Rapamycin reverses cellular phenotypes and

enhances mutant protein clearance in Hutchinson–Gilford Progeria syndrome

cells. Sci. Trans. Med. 2011;3:89ra58.

[10] Sullivan T, Escalante-Alcalde D, Bhatt H, et al. Loss of A-type lamin expression

compromises nuclear envelope integrity leading to muscular dystrophy. J Cell

Biol 1999;147:913–20.

[11] Scaffidi P, Misteli T. Lamin A-dependent nuclear defects in human aging.

Science 2006;312:1059–63.

[12] Heestand BN, Shen Y, Liu W, et al. Dietary restriction induced longevity is

mediated by nuclear receptor NHR-62 in Caenorhabditis elegans. PLoS Genet

2013;9:e1003651.

[13] Harrison DE, Strong R, Sharp ZD, et al. Rapamycin fed late in life extends

lifespan in genetically heterogeneous mice. Nature 2009;460:392–5.

[14] Bjedov I, Toivonen JM, Kerr F, et al. Mechanisms of life span extension by

rapamycin in the fruit fly Drosophila melanogaster. Cell Metab 2010;11:35–46.

[15] Wullschleger S, Loewith R, Hall M. TOR signaling in growth and metabolism.

Cell 2006;124:471–84.

[16] Hanu N, Ra˘chita˘ M. Cell culture procedure and the optometric method to study

cell "in vitro". Rom Arch Microbiol Immunol 1999;58:147–55.

[17] Ra˘chita˘ M, Diaconeasa G. Hyperglycemia induced different responses to stress

in vitro depending on the cells' age. Biogerontology 2002;3(Suppl. 1):1.

[18] Diaconeasa G, Răchită M. Cercetpri privind mecanismele stresului celular intr-un

model experimental original. Anal Univ Bucuresti 2003.

[19] Rosania GR, Chang Y-T, Perez O, et al. Myoseverin, a microtubule-binding

molecule with novel cellular effects. Nat Biotechnol 2000;18:304–8.

[20] Saraiya M, Nasser R, Zeng Y, et al. Reversine enhances generation of

progenitor-like cells by dedifferentiation of annulus fibrous cells. Tissue Eng

Part A 2010;16:1443–55.

[21] Sachs DL, Kang S, Hammerberg C, et al. Topical fluorouracil for actinic keratoses and photoaging: a clinical and molecular analysis. Arch Dermatol 2009;145:659–66.

[22] Mahmoudi S, Brunet A. Aging and reprogramming: a two-way street. Curr Opin Cell Biol 2012;24:744–56.

[23] Minamino T, Komuro I. Role of telomeres in vascular senescence. Front Biosci 2008;13:2971–9.

[24] Larson DW. The paradox of great longevity in a short-lived tree species. Exp Gerontol 2001;36:651–73.

[25] Villeda SA, Luo J, Mosher KI, et al. The ageing systemic milieu negatively regulates neurogenesis and cognitive function. Nature 2011;477:90–4.

[26] Boselli M, Rock J, Unal E, Levine SS, Amon A. Effects of age on meiosis in budding yeast. Dev Cell 2009;16:844–55.

[27] Moskalev AA, Shaposhnikov MV, Plyusnina EN, et al. The role of DNA damage and repair in aging through the prism of Koch-like criteria. Ageing Res Rev 2013;12:661–84.

[28] Garza LA, Yang CC, Zhao T, et al. Bald scalp in men with androgenetic alopecia retains hair follicle stem cells but lacks CD200-rich and CD34-positive hair follicle progenitor cells. J Clin Invest 2011;121:613–22.

[29] Yang C, Tibbitt MW, Basta L, Anseth KS. Mechanical memory and dosing

influence stem cell fate. Nat Mater 2014;13:645–52.

[30] Loffredo FS, Steinhauser ML, Jay SM, et al. Growth differentiation factor 11 is a

circulating factor that reverses age-related cardiac hypertrophy. Cell

2013;153:828–39.

[31] Ito M, Yang Z, Andl T, et al. Wnt-dependent de novo hair follicle regeneration

in adult mouse skin after wounding. Nature 2007;447:316–20.

[32] Partridge L. Evolutionary theories of ageing applied to long-lived organisms.

Exp Gerontol 2001;36:641–50.

[33] Diaconeasa AG. Civilizatia foametei/O alta˘ abordare a

umaniza˘rii. Craiova: Sitech; 2007. p. 124.

[34] Seluanov A, Hine C, Azpurua J, et al. Hypersensitivity to contact inhibition

provides a clue to cancer resistance of naked mole-rat. Proc Natl Acad Sci USA

2009;106:19352–7.

[35] Ogden CL, Fryar CD, Carroll MD, Flegal KM. Mean body weight, height, and

body mass index, United States 1960–2002. Advance data from vital and

health statistics, CDC division of health and nutrition examination surveys.

Adv Data 2004;347:1–17.

[36] Diaconeasa AG. Civilizatia foametei/O alta˘ abordare a

umaniza˘rii. Craiova: Sitech; 2007. p. 89–90

[37] Rohme D. Evidence for a relationship between longevity of mammalian

species and life spans of normal fibroblasts in vitro and erythrocytes in vivo.

Proc Natl Acad Sci USA 1981;78:5009–13.

[38] Crews DE, Stewart JA. Human longevity and senescence. In: Muehlenbein MP,

editor. Human evolutionary biology. Cambridge University Press; 2010. p. 532.

[39] Horvath S. DNA methylation age of human tissues and cell types. Genome Biol

2013;14:R115.

[40] Stephens C. Senescence: even bacteria get old. Curr Biol 2005;15:R308–10.

[41] Wang P, Robert L, Pelletier J, et al. Robust growth of Escherichia coli. Curr Biol

2010;20:1099–103.

[42] Stewart EJ, Madden R, Paul G, Taddei F. Aging and death in an organism that

reproduces by morphologically symmetric division. PLoS Biol 2005;3:e45.

[43] Lindner AB, Madden R, Demarez A, Stewart EJ, Taddei F. Asymmetric

segregation of protein aggregates is associated with cellular aging and

rejuvenation. Proc Natl Acad Sci USA 2008;105:3076–81.

[44] Coquel AS, Jacob JP, Primet M, et al. Localization of protein aggregation in

Escherichia coli is governed by diffusion and nucleoid macromolecular

crowding effect. PLoS Comput Biol 2013;9:e1003038. http://dx.doi.org/ 10.1371/journal.pcbi.1003038.

[45] Coelho M, Dereli A, Haese A, Kuhn S, Malinovska L, De Santis ME, et al. Fission

yeast does not age under favorable conditions, but does so after stress. Curr
Biol 2013;23:1844–52.
[46] Jaskelioff M, Muller FL, Paik J-H, et al. Telomerase reactivation reverses tissue
degeneration in aged telomerase-deficient mice. Nature 2011;469:102–6.
[47] Gomes AP, Price NL, Ling AJY, et al. Declining NAD+ induces a pseudohypoxic
state disrupting nuclear-mitochondrial communication during aging. Cell
2013;155:1624–38

ARE THERE DIFFERENT KINDS OF AGING?

Introduction

If we are to understand aging, our first look should be at syndromes of accelerated, premature aging. Some of these, under the generic term of progeria, are extremely rare diseases during which aging occurs at accelerated rates, with onset in childhood. Patients with these syndromes of premature

aging rarely reach the adult age. Werner syndrome and the Hutchinson-Gilford progeria have been the moststudied progeroid syndromes. The word "progeria" comes from the Greeek words "pro" (πρό), "ahead" or "premature" and "gēras" ((γῆρας) "advanced age".

There are also diseases during which aging is more accelerated, but patients reach adult age and even grow older than that. In this sense, we have noted Down syndrome, a relatively frequent genetic disease, but also type 2 diabetes, which is associated with the degenerations of various organs

and systems. These pieces of information are background knowledge among gerontologists.

In the present work, we attempt to analyze existing data in specialized literature regarding accelerated aging under the generic term "Progeria", aiming to identify several mechanisms which might be common with "normal" aging. Also, we aim to analyze these syndromes of accelerated aging in accordance to the biochemical hypothesis on aging [1]. According to the aforementioned hypothesis, aging would be the result of alterations of specific chemical reactions that occur as a consequence of diminution of cellular signals involved in these reactions. The ordinary functioning of the organism and its evolutionary and ecological variations lead to different alterations depending on the species and environment, reflected by various phenotypes and aging rates.

Specific cellular reactions such as those involved in cellular differentiation, but also those related to the repair of genetic material and protein processing could take place at increasingly lower rates. As a consequence of the aforementioned phenomenon, the rates at which reactions involved in

energy production and storage take place could decrease, with effects on subsequent reactions. Factors of impact on cell energy, such as number and function of mitochondria and other reactions involved in the metabolism of glucose and energy storage compounds, could have effects on the capacity for cellular synthesis and thus would favor aging.

In other words, aging could express itself in two ways: 1. As affecting cell differentiation; 2. As affecting cell energy. These two ways should influence each other. The phenomenon initiating aging should be, according to this hypothesis, the decrease of cellular signals. In the following sections, we will attempt to analyze the syndromes of accelerated aging, taking into account cellular differentiation and cell energy.

Intriguing facts about progeria

The accelerated aging syndromes are genetic diseases most often caused by mutations that affect the encoding of enzymes involved in the repair and structuring of the genetic material, thus affecting the integrity of the genome [2]. The most well-known diseases of this type are Werner syndrome, the Cockayne syndrome, xeroderma pigmentosum, and ataxia telangiectasia, which are all caused by genetic alterations of certain enzymes. Thus, their mechanisms have common traits with

Werner syndrome, and we will discuss it as representing a class of aging disorders.

Werner syndrome, also known as "adult progeria", is an autosomal recessive disease with a frequency of 10 in a million, though more frequent in certain populations. The gene involved in this syndrome encodes a helicase, Werner syndrome ATP-dependent helicase, also known as type 3

RecQ-like, which is an important enzyme for DNA structuring, with roles in DNA repair, but also in replication and recombination, and probably also in telomere maintenance [3]. Subjects with Werner syndrome are born normal but the syndrome starts to manifest itself in the second decade of life, when there is an onset of diseases associated with aging such as cardiovascular, osteoporosis, cancer of the type of mesenchymal neoplasia (sarcomas), alopecia, type 2 diabetes, cataract, hypogonadism. Subject's death generally occurs in the fourth decade of life due to cardiovascular diseases or cancer. It is interesting that fibroblasts removed from patients with Werner syndrome and studied "*in vitro*" afterwards are characterized by an accelerated senescence, meaning that these fibroblasts have a more reduced capacity to divide in cell cultures than those removed from healthy subjects.

The Werner syndrome ATP-dependent helicase, or the WRN protein (an enzyme involved in DNA elongation in view of transcription and other biochemical processes), could have complex cellular functions, being related also with p53, which, due to its involvement in cell division regulation, affects the risk of malignancy [4]. Also, recent data have shown that in the case of Werner syndrome there is an alteration of heterochromatin (non-information DNA sequence) [5]. The differentiation of WRN-null human embryonic stem cells to mesenchymal stem cells recapitulates features of premature cellular aging, specific patterns of methylation associated to cellular aging, and changes in heterochromatin architecture, a phenomenon that also occurs in physiologically aging mesenchymal stem cells. The WRN protein, analyzed in the study by Zhang, could play the role of preserving heterochromatin's integrity, which is crucial in aging.

These data are in accordance with those on physiological aging to which various patterns of methylation were related (epigenetic changes with effects on gene expression) [1]. Nevertheless, the loss of telomeres itself, in the case of cellular senescence, is a loss of heterochromatin. This phenomenon could be part of a broader view on heterochromatin's disintegration, at least for "*in vitro*" cellular aging. Supplementary amounts of an enzyme that is involved in histone methylation prevent losses of heterochromatin in WRN-null stem cells [5]. Hence, methylation might protect heterochromatin. On the other hand, methylation reactions could reflect the cell's energy state. During the gestation period, low intake of folic acid as a source of methylation is responsible for the risk of cardiovascular and degenerative diseases. The level of homocysteine, a metabolite that reflects folic acid deficit, is increased in cardiovascular diseases, Alzheimer's and psychiatric disorders. But attempts to reduce levels of homocysteine (the aforementioned metabolite also having pro-oxidant effects, by supplementations with vitamins B6, B12 and folic acid) did not lead to a decrease in the number of major cardiovascular events [7]. Nonetheless, homocysteine results from methylations of DNA and histones (with likely formation of heterochromatin) [8]. Homocysteine might be re-methylated to methionine by folic acid through the methylenetetrahydrofolate reductase MTHFR pathway. « Since the HGPS mutation occurs *de novo* in a newly conceived zygote when folate-dependence is at its peak, we propose that folate deficiency might be associated with the mutation, but there are no reports demonstrating this association so far » [8].

Methylenetetrahydrofolate reductase is dependent on NAD(P)H. The increased expression of the methyl transferase and increased DNA methylation lead to an equivalent (large) amount of homocysteine.

NAD has an interesting action in aging. Under the influence of mitogenic factors, NAD levels increase in cells removed from young subjects, but not in those from old ones. This difference is not evident in patients of the same age, but with diabetes and Down syndrome. On the other hand, increases in NAD levels with age have been observed, which could be related to the constant protein/NAD ratio of cells that is also higher in old cells [9]. In aging, responses to environmental factors (stress) might be diminished, whereas what changes with age is the capacity to store and use cellular energy. Also, the cellular NAD levels are much higher in patients with progeria [9]. The link between NAD and sirtuin, involved in the maintenance of DNA architecture and prevention of recombination, is much debated in the literature on aging. "Sirtuins" SIRT are a family of protein deacetylases/ADP ribosyltransferases dependent on NAD. Sirtuin deacetylation activities are similar to methylations, because these epigenetic changes play the same role of inactivating genes. It was demonstrated that SIR2 and its orthologues regulate longevity in yeasts, worms and flies. In mammals, SIRT2, SIRT1 orthologues coordinate metabolic reactions as a response to the presence of available nutrients in various tissues [10]. Moreover, it appears that both NAD and a limiting enzyme for NAD synthesis might have circadian variations that in turn influence the transcription of genes whose expression is also circadian. This transcription is mediated by SIRT1, whose role has already been demonstrated. Laboratory rats subjected to caloric restriction had higher WRN levels, while SIRT1 levels are simultaneously increased under caloric restriction. In mice lacking sirt-1, the WRN levels are reduced. These results showed that deacetylation by sirtuins determines constant WRN levels [11].

Although the relation between cell NAD and aging and progeria is based on older studies, and the technology for measuring NAD levels even now is controversial and no recent evidence suggests that NAD levels increase with age or in progeria, other data suggest a relation between cell energy and the risk of mutations. Moreover, in the disease model of Cockayne syndrome, a lipid- and NAD- rich diet changes the transcriptomic, metabolomic and behavior phenotypes of mice [12]. Cockayne syndrome is a syndrome of accelerated aging characterized by neurodegeneration, and in which enzymes involved in DNA repair are affected. Administration of a NAD precursor such as nicotinamide riboside in 24 months old mice (very old) led to a good muscle regeneration, but also to an increase in lifespan [13].

Both lipids (fatty acids) and NAD are reactants for the mitochondrial oxidative phosphorylation. Actually, these reactions stimulate the most important source of cellular energy, namely mitochondrial respiration. Some of the aforementioned reactions are weak in aging, which could lead to a decrease

in the levels of specific cellular products that intervene in cellular differentiation [1].

These data support the idea that there is a deficit of cellular energy in accelerated aging syndromes, which is reflected in both cellular synthesis and heterochromatin maintenance, both in Werner syndrome and Cockayne syndrome. This deficit has implications on transcription, the capacity to differentiate, and cell survival. The "Werner syndrome ATPdependent helicase" -WRN protein itself is dependent on ATP and, as such, dependent on cellular energy.

Folic acid is implicated in the synthesis of macroergic phosphate compounds. Perhaps, the organism adapts itself with age to energy deficits, and thus there is a reduction in DNA methylation and protection of heterochromatin, which maintains its architecture through other methylations. These later mentioned processes consume cellular energy and hence aging would result from adapting to a cellular

energy deficit at several levels. It is noteworthy as well that hyperhomocysteinemia is associated with progeria [8].

Werner syndrome is an accelerated aging in which changes related with aging at DNA level occur earlier than normal. These data are in accordance with our biochemical hypothesis on aging. According to this hypothesis, in "normal" aging and as a consequence of diminished cellular signaling, epigenetic changes are likely to occur, which subsequently leads to changes of heterochromatin. Although the

cellular energy deficit is evident in Werner syndrome, it is interesting whether the same diminished signaling plays a role also in this syndrome as well.

In Werner syndrome, defective transcription could "catalyze" the above phenomena. Helicases are involved not only in replication, repair, and recombination, but also transcription, which presumes DNA unwinding and elongation. It is interesting that the onset of this syndrome occurs in the second decade of life. Probably not only the accumulation of mutations through defective replication for a sufficiently increased number of times, but also the accumulation of transcription's harmful effects, could contribute to the start of the aging process. Actually, mutations (including recombinations) are noxious because eventually they affect survival and cellular functions, meaning transcription, protein synthesis, and processing. Changes in the organism's functions, accompanied by changes in the functioning of various types of stem cells, occur at the age of onset of Werner syndrome. Development in Werner syndrome has several features in common with aging, namely modifications of reaction rates. During development and maturation, some reactions occur at significantly increased rates, to the detriment of other such reactions. The difference with regard to normal development would be that the latter is much more controlled, and hormonal signals, (for instance sexual hormones), stimulate very specific cellular reactions. Sexual hormones function as growth factors. Presumably, in patients with Werner syndrome there would be a transcription deficit that cannot be overcome as it occurs during normal adolescence, at least for some cell types, and hence this deficit in transcription manifests as aging. Actually, what is striking in patients with Werner syndrome is that the "adolescent growth spurt" is not noticeable and this fact leads to diagnosing the syndrome [14]. Possibly, in this case aging might start subsequent to late actions of the sexual hormones involved in cellular energy.

Could defective transcription be the main factor in the onset of Werner syndrome? According to the biochemical hypothesis on aging, defective transcription has similar effects with diminished signaling, which could also be causing the start of aging. This prediction can be illustrated experimentally through alterations of transcription factors, but also by stimulating specific cellular reactions. In fact, one single transcription factor FOXN1 is involved in thymus regeneration [15], and decline in thymus FOXN1 levels is correlated with age [16].

However, the genetic defect in the most spectacular form of progeria, the Hutchinson-Gilford Progeria Syndrome (HGPS), (also called "the child's progeria", in which the aging rate is 8-10 times higher than normal), is expressed by a protein of the nuclear membrane known as lamin A [17]. Lamin A and lamin B are the main proteins forming the nuclear lamina, a filamentous meshwork (proteins of 20-50 nm in mammals), that functions as the interface between the inner nuclear membrane and chromatin [18]. A point mutation of the gene encoding lamin A leads to the defective processing of this protein, (the accumulation of lamin A in the nuclear membrane), which determines blebbings [17]. Since 2003, several articles have confirmed that LMNA is involved in HGPS, described the cryptic splicing site activation due to the LMNA main G608G mutation, which leads to several disorders in DNA replication, transcription, repair, as well as in nucleocytoplasmic transports. All these contribute to defects in nuclear genome metabolic pathways resulting in cell aging

It is interesting that in the case of the Hutchinson-Gilford progeria, the risk for cancer and senile dementia is not increased as it is in normal aging and Werner syndrome. Although patients with this rare syndrome hardly reach the age of twenty years old, their intelligence quotient is slightly higher than the average [19]. This syndrome is extremely rare. At birth children are apparently normal, but with a thinner skin and more visible blood vessels. In the months following birth, these children have some symptoms, among which no weight gain, localized scleroderma-like symptoms meaning skin thickening, and creasing especially to the extremities. Between the age of 18 and 24 months, specific changes such as slow-growing, generalized alopecia and a specific look with a small face, a retracted maxilla and a convex nasal profile appear. As the child grows older, she or he exhibits signs of aging and diseases associated with aging. The skin wrinkles and cardiovascular diseases, atherosclerosis, loss of vision, and renal failure occur. After that, muscle, skeleton degeneration, hip dislocation and fat loss occur, and joints are rigid. Generally, the death of patients with HGPS is caused by cardiovascular complications.

The relations between the blebbings of the nuclear membrane and changes similar to aging that the former mentioned determine is not presently known. Moreover, it appears that lamin A also accumulates during normal aging, though at lower rates than during progeria [20].

Lamin A production in large amounts is activated in „*in vitro*" senescent cells. It is interesting that some studies have suggested that lamin A might be responsible for changes of the genetic material which are associated with aging, such as DNA defects and histone methylations (that in turn have influence on heterochromatin) [21]. Histone-covalent modifications are radically different from control specimens, with the tendency to lose the bipartition between euchromatin and heterochromatin. This is reflected in an altered spatial compartmentalization and conformation of chromatin within the nucleus. However, it is noteworthy that in patients with Hutchinson-Gilford progeria, epigenetic changes typical of "normal" physiological aging did not occur [22].

Could lamin A be the main cause of phenomena in HGPS? Lamin A is related to the nucleo-skeleton [18], and this fact might affect transcription. Under these conditions, a conclusion could be drawn that transcription reactions could be limited because of signalling at the level of nucleus. Abnormal interactions of nuclear lamina's proteins could lead to the phenotypes of progeroid syndromes and especially progeria. Probably, the nuclear membrane's blebbings prevent hormones and nuclear receptors from acting in view of signalling transcription. In the case of Hutchinson-Gilford progeria, overall transcription could be affected in contrast to „normal" aging.

What could be the cause of Hutchinson-Gilford's mutation?

In the history of studying the above syndrome HGPS was initially considered to be a recessive autosomal disorder, as it appeared in a larger proportion in families with consangvinization [19]. Genetic studies showed that 90 % percent of patients with HGPS had a point mutation, which subsequently appeared in some patients as a "de novo" mutation through the paternal line.

Sporadic mutations of paternal alleles occur also in other syndromes, such as achondroplasia, Apert, Crouzon, and Pfeiffer syndromes, and were attributed to fathers' advanced ages (same with Hutchinson-Gilford progeria). The advanced ages of fathers could be correlated with higher frequencies of mutations in sperm's genetic material. Nevertheless, in the case of achondroplasia, the above hypothesis was rejected [18].

The mutation responsible for Hutchinson-Gilford progeria is a cytosine-thymine substitution at codone 608 in a cytosine guanine CG dinucleotide -rich area (isochore) in which methylations are frequent. It is even considered that a methylated cytosine could easily undergo deamination followed by an erroneous transcription. So, these reactions would take place in an area predisposed to mutations, but also epigenetic changes, which, in our opinion, could subsequently be the source for other mutations.

LMNA point mutations also occur in other progeroid syndromes, for instance Wiedemann–Rautenstrauch, which is characterized by severe lipodystrophy and various base substitutions, which preponderantly affect various tissues (muscle) [23].

The mutation that occurs in Hutchinson-Gilford progeria and affects the pre-lamin A protein prevents the removal of a farnesyl group (aliphatic rest as a cholesterol precursor). In absence of the above post-translational change, an abnormal lamin A, namely progerin, permanently attaches to the nuclear rim. Progerin leads to abnormal formation of the nuclear lamin. In absence of this mutation, the farnesyl group is removed from the normal lamin A protein, which reaches the nuclear envelope. These processes do not occur in the case of progerin, which is just lamin A protein from which the farnesyl group was not removed.

Drugs used in an attempt to treat Hutchinson-Gilford, which concentrated on the farnesylation reaction, were those utilized in cancer, such as lonafarnib [24] (which was also subject of a clinical trial conducted on patients) [25], but also statins [26] or zolendronat, used to treat hypercalcemia. The results of the clinical study on lonafarnib showed some moderate benefits.

If the mutant variant of lamin A protein also occurs in "*in vitro*" cultivated cells from young and healthy subjects, after several successive multiplications [20], while lamin A accumulates in the nucleus in old healthy subjects without an increase in mRNA, the cause could be attributed to the defective protein processing responsible for the aforementioned accumulation. In other words, we can presume that abnormal processing of other proteins might lead to this phenotype of aging. The good question, if we take into account this hypothesis, would be whether there are subjects carrying this mutation but not developing the syndrome, or healthy siblings of patients with the HGPS syndrome who carry this mutation.

Facts were further complicated when LMNA mutations were discovered in the heterozygote parents of a child with progeria. The parents also had symptoms of other progeroid syndromes [27]. This child's parents, who were carriers of different mutations, had 6% fibroblasts with abnormal nuclei. Abnormalities of the nucleus both in the child (36% abnormalities of the nuclei) and his parents were different from those taken into account as classical (nuclear membrane's blebbings) for Hutchinson-Gilford progeria. In this case, abnormal lamin A was not accumulated in the nuclear periphery, but was leading to other nuclear abnormalities, shapes of honeycomb and blebbings in which lamin B (another protein of the nuclear envelope) was lacking. More than that, inhibitors of farnesylation did not lead to the repair of the cellular phenotype as in the case of the most common mutation in Hutchinson-Gilford progeria, but to other nuclear abnormalities including accumulation of pre-lamin A to the peripheral nuclear lamin.

These results showed not only that the terminal C area of pre-lamin A is very important for the function, but also that the accumulation of this protein would not be the decisive harmful factor for the symptoms [27].

But the most interesting fact is that there are patients with both HGPS and Werner syndrome who do not have any of the mutations mentioned above (approximately 10%). However, a study showed that all patients with these syndromes had various methylation patterns compared with age matched healthy subjects. Both in HGPS and Werner syndrome methylation patterns are different in case of present mutations from those in case of absent mutations [28].

What is more interesting, a case was reported for a healthy subject with the same methylation pattern with that of children having HGPS, but no LMNA mutation. The case was that of the father of two little girls with HGPS [27]. Although new mutations of LMNA involved in progeroid syndromes were discovered, some of the patients with progeroid syndromes and other accelerated aging syndromes have no known mutation.

In this context we may suggest that a certain pattern of cellular signaling might lead to a certain methylation pattern (epigenetic changes), which in some cases could determine "de novo" mutations in CG-rich areas. This is how the LMNA mutation present in a large number of patients with HGPS (90%) might occur as a very likely phenomenon in the case of these epigenetic changes. In what case would the aforementioned epigenetic changes be investigated?

Folate deficiency, as a symptom of cell energy deficit, may lead to a higher rate of general mutations, as suggested by the relation between this deficiency and cancer [6]. The mutation involved in HGPS is a very rare one, so its likelihood is very low, but it is also possible among other mutations. The symptoms of these syndromes are the answer.

Low cell energy and fibrosis

In the case of HGPS, the majority of symptoms can be considered as related to extended fibrosis. Fibrosis is at the origin of most symptoms of the HGPS and implicated in heart diseases such as atherosclerosis. It was assumed that fibroblasts from patients with HGPS progeria undergo hyperproliferation followed by apoptosis [18]. But more recent papers evidenced the apoptosis of smooth muscle cells (SMC) from the vascular bed, the replacement of SMC by fibrotic tissue, both being the main cause of vascular defects leading to the death of HGPS patients from strokes or heart attacks. Concordantly with this, progerin accumulation reduces the proliferative and differentiation capacity of pluripotent and multipotent mouse and human cells [29].

Probably the above processes take place in the same way in normal aging, at least in some tissues and at some points in time. Fibrosis is characterized by few cells and many fibers. Lamin A could be involved in replication, transcription and DNA repair. Through interactions with retinoblastoma pRb proteins, lamin A would be implicated in cell proliferation and differentiation [27]. Proteins of the the type of lamin A play roles in signaling pathways involving beta-1 tumor growth factor (TGF beta-1), and these processes could explain the relationships between a cellular accelerated turnover and premature aging. Lamins of type A could modulate production of collagen through TGF beta-1 induced signaling.

In lamin A protein knockout mice, the production of collagen is increased and this fact would explain increased fibrosis in murine animals of experimental models of Hutchinson-Gilford progeria and patients with laminopathies [27].

But fibrosis (hypersecretion of collagen as associated with few cells) also occurs under conditions in which LMNA is absent, these conditions being often associated with "normal" aging.

It is well-known that areas with high concentrations of proteic fibers are very affected by aging. For example, joints tear very fast. Free radicals attack, glycosylations and crosslinking are enumerated as mechanisms involved in joint degenerations, as numerous data have shown that the aforementioned reactions lead to the degradation and loss of functions of proteins. But a rapid protein turn-over, as is characteristic of young age, could prevent the degradation process.

Caloric restriction, which increases average and maximum life-spans in many phylogenetically distant species, also increases protein turn-over and autophagy. In fact, in patients with HGPS, rapamycin slowed cellular senescence, increased progerin degradation in these cells and restored blebbings of the nuclear membrane. The formation of insoluble aggregates of progerin also decreased under rapamycin, which induces clearance through autophagic mechanisms in normal fibroblasts [20]. These have been in support of the hypothesis according to which enzymes involved in protein degradation are affected by progeria but also “normal” aging.

A conclusion we may draw is that proteins involved in structural support, both at the organismal level and the cytoskeleton play a very important role in aging. Meshworks of protein fibers not only provide (mechanical) support but also have important interactions with proteins involved in cellular signaling, and consequently in cellular proliferation, differentiation and functioning. Likewise, protein processing is a very complex process which requires many enzymes of various cellular compartments and reactions may lead to numerous errors. It is known that, in aging, post-translational processing of proteins may be defective. Abnormal glycosylations of some membrane proteins can occur and thus affect proteins' functions or determine autoimmune reactions.

We believe that in the case of progeria and progeroid syndromes, aside from mutations in LMNA, it is very likely that other proteins which are involved with protein processing, cell proliferation and differentiation might play important roles. If there are subjects with mutations in LMNA located at key sites but nonetheless healthy, the above assertion about important roles of other proteins is endorsed. The fact that similar cellular phenotypes to that of progeria have been shown in normal aging and as well in senescent cells from young patients, also supports the above assertion. Also, the fact that the progeria was found out in families with consangvinization may suggest a lack of alternative enzymes (genes) to compensate for the deficient reactions. For such a phenotype, the low probability to be discovered might result from the fact that, aside from mutations whose probabilities to occur are low, other factors and other enzymatic deficiencies could contribute to the above mentioned phenotype.

A small number of cells mean a low capacity to divide and a high level of collagen means an increased cell differentiation. Taking into account that an upregulation of the insulin/IGF1 (insulin growth factor 1) signaling pathway is associated with cell proliferation, and a downregulation with cell differentiation [6], fibrosis could indicate a cellular energy deficit.. Recent data have suggested that cellular energy is indeed affected in HGPS. The question would be whether mitochondrial functions being affected by defective lamin A synthesis or mitochondrial dysfunction is related to an increased expression of progerin. In fibroblasts from patients with Hutchinson-Gilford syndrome, a downregulation of mitochondrial oxidative phosphorylation proteins accompanied by mitochondrial dysfunction was observed. Mitochondrial dysfunctions were also found in fibroblasts from adult progeroid mice expressing progerin (LmnaG609G/G609G knocking mice) or prelamin A (Zmpste24-null mice). The analysis of tissues from these mouse models revealed that the damaging effect of these proteins on mitochondrial function is time- and dose-dependent. Mitochondrial alterations were not observed in the brain, a tissue with extremely low progerin expression that seems to be unaffected in HGPS.

Mitochondrial function was restored in progeroid mouse fibroblasts treated with the isoprenylation inhibitors FTI-277, or pravastatin plus zoledronate, which are being tested in HGPS clinical trials. These results suggest that mitochondrial dysfunction contributes to premature organ decline and aging in HGPS. Beyond its effects on progeria, prelamin A and progerin may also contribute to mitochondrial dysfunction and organ damage during normal aging, since these proteins are expressed in cells and tissues from non-HGPS individuals, most prominently at advanced ages [30]. Other data showed that at sites which are affected by fibrosis such as the cartilage from patients with Osteoarthritis (OA) expression of lamin A was markedly elevated compared with non-OA control samples [31].

Methylene-blue, which is an antioxidant with positive effects on mitochondria (there is a large proportion of abnormal mitochondria in

HGPS), helping their survival, not only has anti-apoptotic effects on cells removed from patients with HGPS, but also restores the nucleus' shape [32]. Other analyses regarding methylene blue suggest that it induces the release of progerin from the nuclear membrane, protects perinuclear heterochromatin and restores defective expression of genes isolated from cells of HGPS patients.

These data also suggested that the cellular energy deficit would be the factor leading to abnormalities of protein synthesis and processing, so abnormal proteins such as progerin are produced. This idea is supported by the fact that mitochondrial dysfunction has been linked to the pathogenesis of progeroid syndromes such as HGPS and Cockaine syndrome. Also, cells derived from HGPS patients show a marked downregulation of mitochondrial oxidative phosphorylation proteins, accompanied by mitochondrial dysfunction [18]. If the cellular energy deficit that occurs in aging as a result of diminished signaling is also involved in progeria, then some metabolic pathways such as IGF1/insulin are affected.

In fact, in models using laboratory mice there is a change of the ratio IGF1/GH (growth hormone). Also, IGF1 delayed progeria's effects in mice [33], but these results are not valid in humans. Probably in humans the IGF1 levels are much lower than in mice, so effects of this hormone would not be striking in comparison with small mammals. There is an inverse relationship between a mammal's body weight and IGF1 levels. IGF1 intervenes in cell survival and proliferation. Presumably in humans there are also effects, but more subtle. In any case, a complex alteration of the IGF1/insulin signalling may be involved in aging and progeria. In fact, progeroid syndromes seem to be accompanied by metabolic changes. Zmpste24−/− mice exhibit profound transcriptional alterations in circulating levels of GH and IGF-1, as well as changes in metabolic pathways that are associated with autophagy induction. Metabolic alterations are also present in progeroid syndromes that feature defects in DNA repair.

Wrn-deficient mice exhibit hypertriglyceridemia and insulin resistance. Furthermore, mice with telomere dysfunction exhibit a marked deficiency in IGF-I-mTOR signalling,impaired energy homeostasis, and suppressed mitochondrial biogenesis [29]. But what is the original cause: metabolic impairment or DNA defects? A study on mice with alternative splicing of mRNA for LMNA shows that mice with lamin C live longer, have decreased energy metabolism, increased weight gain, and reduced respiration. In contrast, progerin-expressing mice show increased energy metabolism and are lipodystrophic. Increased mitochondrial biogenesis is found in adipose tissue from HGPS-like mice, whereas lamin C-only mice have fewer mitochondria [35]. Progerin seems to have aging effects per se, but it may appear in special metabolic conditions.

In HGPS, the sequence of phenomena could be the following: certain epigenetic changes as a consequence of a modified metabolism that is characterized by cellular energy deficit, favor mutations such as the LMNA. The energy deficit further manifests itself by affecting mitochondria and having as consequences apoptosis, but especially fibrosis. Children with HGPS have at birth thin skin and visible blood vessels, these aforementioned characteristics showing deficient tissue development. In may be the case that in these children metabolic energy transitions from the embryonic period to the neonatal days fail. Cells from these children patients extensively and prematurely differentiate, instead of proliferating, and hence, fibrosis occurs. In an experimental model, a hypersecretion of collagen occurs as well in normal cultivated keratinocytes after a certain number of passages [34]. The above phenomena could explain the low incidence of cancer, in which there is an insulin/IGF1 hypersignaling

[6].

Table 1. Possible common and different traits between the accelerated aging syndromes and "normal"aging.

disease	Mutation affecting DNA repair and genome integrity	Decrease of cell energy and capacity for protein synthesis	Affected cell signaling/cell differentiation
HGPS	possible	possible	x
Werner syndrome	x	possible	x
Cockayne syndrome	x	possible	x
xeroderma pigmentosum	x	possible	x
ataxia telangiectasia	x	possible	x
diabetes mellitus	no	x	x
Down syndrome	no	possible	x
Normal aging	x	x	x

As a conclusion, Werner syndrome, Cockayne syndrome, xeroderma pigmentosum, ataxia telangiectasia are associated with impaired cell differentiation; that is the reason they increase the risk of cancer. Other accelerated aging syndromes, especially HGPS, are not. Looking at the table above, we can infer that these syndromes just accentuate some traits of "normal" aging.

Conclusions

By analyzing data with regard to syndromes of accelerated aging, which all are under the generic term of "progeria", we advance the following perspective on the topic. We consider that progeria and other progeroid syndromes are a good such model for aging, keeping its limitations in mind. Patients with progeria differ from old patients for several reasons, not only because of pathologies, which do not totally overlap in these two categories of subjects.

Patients with progeria already develop degenerative pathologies in childhood. Hence, in their case, the mature stem cell proliferation as well as the cellular differentiation stop. In the case of "normal aging", there is a different starting point, the regenerative capacity of the organism being higher when

the aging-associated biological decline occurs. In this sense, progeria would be comparable with aging at very advanced ages.

Also, by comparing "normal" aging with progeria, we can differentiate degenerative processes such as defective regeneration and abnormal regeneration occurring in cancer and dysplasia and relate them to repair processes, in which there are errors at all levels. We consider that in HGPS, the aforementioned defective repair processes are rare, and this fact is an explanation for reduced frequencies of cancers and neurodegenerative processes.

It is also interesting that Hutchinson-Gilford progeria has a rapid progression, death also occurring earlier than in other syndromes of accelerated aging such as Werner syndrome. These data have suggested that mutations are not the most important factor in aging, in the sense that it is not the increased mutation rate that lead to the most accelerated rates of aging, but other biochemical factors with direct impact on protein synthesis and processing.

Taking into account the biochemical hypothesis of aging, we may draw the conclusion that both in Werner syndrome and Hutchinson-Gilford syndrome, a cellular energy deficit occurs, as well as diminished cellular signaling, the extents to which these two later mentioned occur being nevertheless different. In Werner syndrome, diminished cellular signaling occurs to a larger extent, whereas in

HGPS, the cellular energy deficit would be a most important factor. In this context, the cellular energy deficit seems more life-threatening than cancer and is associated with a more accelerated rate of aging.

Future studies regarding progeria may contribute to uncovering mechanisms of the aging process, but also of the aging-associated diseases. But according to these data and the biochemical hypothesis of aging, as shown in Table **1**, "normal aging" may result from alterations present in the accelerated aging syndromes, but with different proportions.

LIST OF ABREVIATIONS

HGPS = Hutchinson-Gilford Progeria Syndrome

WRN Protein = The "Werner Syndrome ATP-dependent Helicase"

LMNA = The Gene Encoding Lamin A

IGF1 = Insulin Growth Factor 1

OA = Osteoarthritis

GH = Growth Hormone

IGF = 1-mTOR

Parts of this chapter appeared in Diaconeasa, Amalia G., and Mariana Rachita. "Are there Different Kinds of Aging?." *Current Aging Science*11.1 (2018): 55-62.

References

[1] Diaconeasa A G Rachita M, Stefan-van Staden R-I A new hypothesis of aging, Med Hypotheses, 2015; 84: 252-257.

[2] Best BP, Nuclear DNA damage as a direct cause of aging RejuvRes 2009, 12: 199-208.

[3] Multani AS,Chang S, WRN at telomeres: implications for aging and cancer. J Cell Sci 2007; 120: 713-21.

[4] Yang Q, Zhang R, Wang XW, Spillare EA, Linke SP, SubramanianD, *et al.*, The processing of Holliday junctions by BLM and WRNhelicases is regulated by p53". J Biol Chem 2002; 277(35): 31980-7.

[5] Zhang W, Li J, Suzuki K, Qu J, Wang P, Zhou J. *et al*, Aging stemcells. A Werner syndrome stem cell model unveils heterochromatin alterations as a driver of human aging. Science 2015; 348: 1160-3.

[6] Diaconeasa G, Civilizatia foametei/o altă abordare a umanizării, 2007, Sitech, Craiova

[7] Lonn E, Yusuf S, Arnold MJ, Sheridan P, Pogue J, Micks M, *et al.* Heart Outcomes Prevention Evaluation (HOPE) 2 Investigators. Homocysteine lowering with folic acid and B vitamins in vascular disease, New. Engl .J Med 2006; 354: 1567-77.

[8] Bhargava S and Tyagi SC, Nutriepigenetic regulation by folate–homocysteine–methionine axis: A review, Mol Cell Biochem 2014:387; 55-6

[9] Chapman ML, Zaun MR, Gracy RW, Changes in NAD levels in human lymphocytes and fibroblasts during aging and in premature aging syndromes. Mech Ageing Dev 1983; 21: 157-67.

[10] Imai S-I, "Clocks" in the NAD World: NAD as a metabolic oscillator for the regulation of metabolism and aging, BBA-Proteins Proteom 2010; 1804; 1584-1590.

[11] Kahyo T, Mostoslavsky R, Gotoc M, Seto M, Sirtuin-mediated deacetylation pathway stabilizes Werner syndrome protein, FASEB letters 2008; 582: 2479-2483.

[12] Scheibye-Knudsen M, Mitchell SJ, Fang EF, Iyama T, Ward T *et al.* A High-Fat Diet and NAD+ Activate Sirt1 to Rescue Premature Aging in Cockayne Syndrome, Cell Metab 2014; 840-855.

[13] Zhang H, Ryu D, Wu Y, Gariani K, Wang X *et al.* NAD repletion improves mitochondrial and stem cell function and enhances lifespan in mice. Science, 2016 DOI:10.1126/science.aaf2693

[14] Epstein CJ, Martin GM, Schultz AL, Motulsky AG. Werner's syndrome a review of its symptomatology, natural history, pathologic features, genetics and relationship to the natural aging process. Medicine (Baltimore) 1966; 45: 177-221.

[15] Bredenkamp N, Nowell CS, Blackburn CC. Regeneration of the aged thymus by a single transcription factor. Development (Cambridge, England). 2014; 141: 1627-1637.

[16] Reis MD dos S, Csomos K, Dias LPB, *et al.* Decline ofFOXN1 gene expression in human thymus correlates with age: possible epigenetic regulation. Immunity & Ageing: I & A. 2015; 12: 18. doi:10.1186/s12979-015-0045-9.

[17] De Sandre-Giovannoli A, Bernard R, Cau P, Navarro C, Amiel J, Boccaccio I *et al.*, Lamin a truncation in Hutchinson-Gilford progeria,. Science 2003; 300: 2055,

[18] Pollex RL, Hegele RA, Hutchinson–Gilford progeria syndrome., Clin. Genet, 2004: 66: 375-381.

[19] Brown WT, Progeria: a human-disease model of accelerated aging,. Am. J. Clin. Nutr. 1992;55: 1222S–4S.PMID1590260.

[20] Cao K, Graziotto JJ, Blair CD, Mazzulli JR, Erdos MR, Krainc D, *et al.* Rapamycin Reverses Cellular Phenotypes and Enhances Mutant Protein Clearance in Hutchinson-Gilford Progeria Syndrome

Cells, Sci Transl Med.2011; 3: 89ra58.

[21] Scaffidi P, Misteli T, Lamin A-dependent nuclear defects in human aging,. Science, 2006, 312: 1059–63.

[22] Horvath S, DNA methylation age of human tissues and cell types.

Genome Biol 2013; 14(10)

[23] Cao H, Hegele RA, LMNA is mutated in Hutchinson-Gilford progeria (MIM 176670) but not in Wiedemann-Rautenstrauch progeroid syndrome (MIM 264090) .J Hum Genet 2003; 48: 271-4.

[24] Capell BC, Erdos MR, Madigan JP, Fiordalisi JJ, Varga R *et al.* Inhibiting farnesylation of progerin prevents the characteristic nuclear blebbing of Hutchinson-Gilford progeria syndrome Proc. Natl Acad. Sci .USA, 2005, 102: 12879-84.

[25] Ullrich NJ. Kieran M W, and KleinmanM E. Neurologic features of Hutchinson-Gilford progeria syndrome after lonafarnib treatment. Neurology, 2013, 81: 427-30.

[26] Varela I, Pereira S, Ugalde AP Navarro CL, Suárez MF, Cau P *et al.* Combined treatment with statins and aminobisphosphonates extends longevity in a mouse model of human premature aging . Nat.

Med. 2008, 14: 767-72.

[27] Verstraeten VL, Broers JL, van Steensel MA, Zinn-Justin S, Ramaekers FC, *et al.* Compound heterozygosity for mutations in LMNA causes a progeria syndrome without prelamin A accumulation. Hum Mol Genet 2006; 15: 2509-22.

[28] Heyn H, Sebastian Moran S, Esteller M, Aberrant DNA methylation profiles in the premature aging disorders Hutchinson-Gilford Progeria and Werner syndrome, Epigenetics. 2013; 8: 28-33.

[29] Carrero D., Soria-Valles C, López-Otín C, Hallmarks of progeroid syndromes: lessons from mice and reprogrammed cells, Disease Models & Mechanism 2016; 9: 719-735.

[30] Rivera-Torres J, Rebeca Acín-Perez R, Cabezas-Sánchez P, Osorio FG, Gonzalez-Gómez C *et al.* Identification of mitochondrial dysfunction in Hutchinson–Gilford progeria syndrome through use of stable isotope labeling with amino acids in cell culture, J Proteomics 2013; 91:466-477.

[31] Attur M, Ben-Artzi A, Yang Q, Al-Mussawir HE, Worman HJ, *et al.* Perturbation of nuclear lamin A causes cell death in chondrocytes, Arthritis Rheum 2012; 64: 1940-9.

[32] Xiong Z-M, Choi JY, Wang K, ZhangH, Tariq Z *et al.* Methylene blue alleviates nuclear and mitochondrial abnormalities in progeria, Aging Cell 2016; 15: 279-290.

[33] Sadeghi-Nejad A, Demmer L, Growth hormone therapy in progeria. J. Pediatr. Endocrinol. Metab. 2007; 20: 633-7.

[34] Manner G, Cell division and collagen synthesis in cultured fibroblasts, Exp Cell Res 1971; 65; 49-60.

[35] Lopez-Mejia IC, de Toledo M, Chavey C, Lapasset L, Cavelier P,Lopez-Herrera C, Chebli K, *et al*, Antagonistic functions of LMNA isoforms in energy expenditure and lifespan. EMBO Rep, 2014, doi:10.1002/embr.201338126. PMID: 24639560.

MISSING LINKS IN CELL AGING

Cell culturing started to be conducted at the end of the 19-th century. One of the first ideas that were issued subsequent on cultivating cells, was even being related to aging. In 1912 Alexis Carell was cultivating fibroblasts while questioning himself why organisms grow old given that cells are immortal. Cells which he cultivated appeared as immortal because when he was replacing cell culture medium with fetal serum, he was adding new cells to old cultures, hence rejuvenating the culture.

People had been convinced of immortality of normal cells ever since. This happened until 1961, when Hayflick and Moorhead (1), by improving cell cultivating procedure, noticed that cells do not replicate indefinitely and so, a cell culture can be multiplied for only a definite number of times. Beyond that number of passages, normal undifferentiated cells do not further divide and do not respond to mitogen factors. These cells are called senescent or cells in replicative senescence (2). The phenomenon of limitative replication was termed "replication limit" or Hayflick limit (term established by Sir MacFerlaine Burnet, the founder of modern immunology).

The idea that there has to be a link between the replication limit of cultured cells and organism aging, immediately issued next. On the other hand, this breakthrough impelled research in the domain. Numerous studies revealed all sorts of correlations between life span of a species and number of divisions cells isolated from these species, can undergo.

First, more research teams found out an inverse correlation between the age of the donor and the number of divisions at which human fibroblasts enter senescence (3). However, this correlation was not always being found out (4). From an optimistic viewpoint, this correlation may suggest that the replication potential progressively recedes whilst the organism grows old.

The second aspect, inter-species comparisons with limited values showed a positive (direct) correlation between lifespan of a species and the replication potential of fibroblast cultures. The most comprehensive study used cells from eight mammals species (5), in which at least one research included cells from Galapagos turtles, the latter species known for being very longeval and with a negligible senescence (6).These data suggest that there should be superimposing of genes controlling replication lifespan of cell cultures with those controlling lifespan of an organism.

The third aspect is that cells originating from human individuals with premature aging enter the replicative senescence after a number of divisions smaller than that corresponding to those originating from healthy individuals (6). This fact once more suggests superimposing of genes controlling the replicative potential with those controlling the phenotypes of an aging organism.

It is noteworthy that the replication limit only applies to normal undifferentiated cells. Malignant cells divide infinitely in cultures. HeLa cells, for instance, date for decades and multiply countless times in various laboratories worldwide.

There was only one step from this knowledge to hypothesizing about the importance of the replication limit for evolution. Leonard Hayflick who was a scientist in forensics, issued two such hypotheses (7).

The first hypothesis which was issued shortly after the finding of replication limit, assumes that the later would be responsible for the aging process. The second hypothesis refers to the replication limit as occurring during ontogeny to protect organisms against cancer. Hence, there are diverse errors occurrring in time, that can lead to cancer. Because of the aforementioned, nature developed a system to stop aged cells abundant of such errors, to divide.

This hypothesis had also a great impact on the scientific community. From a theoretical point of view, the hypothesis supported with a major argument the antagonistic pleiotropy, which explained aging through activities of genes that are "useful" for the individual's development at a certain point in time, but deleterious later. In this sense, genes protecting an individual from cancer during his young age, determined afterwards, aging of the organism. The above mentioned are interesting facts, but which was the molecular mechanism of the replication limit?

This mechanism was proposed for the first time by the Russian researcher Olovnicov, who assumed, without experimental evidences, that the mechanism has to be related with the replication at the "terminus" of chromosomes (8). These data are dated in the seventies but only in the middle of the nineties, the discovery of the telomerase confirmed the assumption of Olovnicov.

Cellular aging from the viewpoint of the telomeres/telomerase

In 1930 H.J. Muller and Barbara McClintock independently of each other, depicted telomeres as protective structures localized at chromosomes' ends. In Greek, "telos" means end, extremity, and "meros" is part. Telomeres are nucleo-protein complexes made of (TTAGGG)n hexameric units that are non-coding DNA. Telomeres protect chromosomes against degradation, fusion, recombination and contribute to genome stability.

Telomeres cannot be completely replicated and with each cell division telomeres progressively shorten to a critical length at which divisions stop and that is termed „replicative senescence". The later is considered an internal mechanism that is counting for the aging process at cellular level.

Telomerase is an enzyme ribonucleoprotein complex that plays roles in telomeres replication and maintenance of their length. This enzyme was discovered in 1984 by Elisabeth Blackburn (9), but then described in tumor cell lines and at present detected in 90% out of 2600 tumor samples (10). Telomerase contains two units: a) RNA (hTR)- that provides template for synthesis of hexameric units (and functioning as a sort of „primer", so RNA (hTR)- is like a DNA polymerase miniature specialized for telomeres) and b) reverse-transcriptase (hTERT) which is the catalytic unit.

In conclusion, telomerase is an enzyme involved in replication of telomeres that are the chromosomes' ends. The principle of telomerase activity is simple. At each new cell division, the last repetitive non-coding DNA sequences (telomeres) do not replicate due to not being accessible for the enzymes implicated in replication which simply bind at this site. When telomeres reach a certain critical length after a number of divisions, the cell is no longer able to divide. Telomerase contains a RNA which functions as sort of a primer, hence this enzyme is like a DNA polymerase miniature.

Discovery of the telomerase launched new research directions and gave hope. As in most types of normal cells telomerase is absent except for epithelial cells, lymphocytes and endometrial (cells with active mitosis), the enzyme being present in stem and embryonic cells, but mostly in many types of malignant cells, the aforementioned seemed the key to explain the cell replication limit and implicitly aging of the organism, while blocking telomerase could have led to design treatments for cancer (11).

Telomerase was added to normal cell cultures and increased their lifespan without modifications (12). But telomerase is not the key to increasing lifespan of all types of normal cells. Also, regarding cancer, blocking telomerase did not lead to an outcome (13).

Moreover, some tumoral cells isolated from humans are telomerase deficient but have an alternative way (ALT) (14) the mechanism of which is yet unknown (probably a process of DNA recombination), to lengthen their telomeres and divide (15).

Stabilisation of telomeres' (certain) length is endowing with immortalisation but the aforementioned is not sufficient for malignant transformation. The later mentioned involves a multi-step process that includes activation of oncogenes (p53, pRB), inactivation of tumoral suppressor genes and a mitogen uncontrolled stimulation.

Some data show the very important role of retinoblastoma (RB) in nerve regeneration. Nerve cells have mechanisms to block nerve growth through retinoblastoma recently discovered in these cells (16).

The relationship between aging, cancer and telomerase becomes more complicated. Some data suggest nevertheless that telomerase could have anti-aging effects at least in some species. Evidences in this sense are related to the fact that "knock-out" rats for telomerase and then treated with telomerase, not only had a slower aging rate but also effects of premature aging began to dwindle (17). Mice organisms are characterized by a high telomerase level, telomeres from mice are very large and from our viewpoint, also likely to be due to mice's intensive metabolism.

Also there are species (mice, birds) in which there is no loss of telomerase after embryonic development and the above species are not the ones characterized by longevity and unknown aging markers. On the contrary, longevity of species correlates inversely with length of telomeres (18).

As well, in some tissues (skin) there is shortening of telomeres but not in rat brain (19) or human skeletal muscle (20). Some data suggest a shortening of telomeres in accelerated aging associated syndromes (21), but length of telomeres varies among circulating lymphocytes from the same individual, similar cells isolated from individuals of same age or fibroblast cultures from the same individual (22). Some pathological conditions also seem to be associated with changes in telomere length. Nevertheless, the later mentioned plays a very important role in cell division.

Most probably, from our point of view, the telomerase decline is a symptom of cellular aging. Naturally, a restoration of this enzyme's function would appear if cell signalling was supporting a rapid division. Loss of telomerase in certain cell types, but not others, which may be correlated with a definite cell division rate, could represent one of the cell changes with age caused by an alteration of cell signaling. The aforementioned is an idea encompassed by the biochemical hypothesis of aging we developed (23).

But if telomerase actually does not appear to be " the fountain of youth", stem cells, in this sense. a class of cells that contain telomerase, remain a great hope. If the absence of either telomerase activity or another analogue mechanism is the origin for the replication limit of *in vitro* cells, the question remains about what happens to *in vivo* cells whilst organisms grow old?

Knowledge in relation with the aforementioned is least-clear. In this sense, the scientific community is fairly biased. Some authors totally decline the existence of an *in vivo* replication limit and take the later into account as an artifact of cell culturing. Other researchers search for evidences that prove the existence *in vivo* of a replication limit. The link between aging and replication limit has been even less documented.

Anyway, organisms do not live for such a longtime as to replication limit be reached by cells. Not even in the case of oldest humans, cells isolated from them did not undergo a sufficient number of divisions as to reach the replication limit.

Under these circumstances, Hayflick made an amendments to his hypothesis in view of a new one, which at present is fairly well controversial: the replication limit is not related with aging but instead is linked to the maximum of lifespan of a species (24).

Some authors, based on the *in vitro* experiments, consider that in fact, it is not even necessary to reach the maximum number of cell divisions to enter the cellular senescence. At each new cell division, a cell is at risk to become senescent. Noxious environmental factors (ionizing radiations, pro-oxidant conditions) seem to accelerate the senescence process. *In vitro* lifespan of fibroblasts from rats doubled as accounted for by the number of possible cell divisions. Comparison was for example, between fibroblasts when grown under a gas atmosphere of 5% O2 and fibroblasts when grown under a gas atmosphere of 20 % O2.

Another aspect is that according to some data, cells removed from the same individual (biopsies) at a certain time, then added to cell cultures, undergo a different number of divisions and this facts rule out the age of the individual from the equation for replicative senescence. The same is in the case of the number of divisions of fibroblasts isolated from the same individual. At different ages of the individual, the number of fibroblast divisions can be maintained constant (21).

What happens for certain with cells from old living organisms?

From our available data, the immune system is the localization where some specialists suggest that cells enter senescence. This phenomenon would determine reduced resistance to infections of old individuals. Memory T lymphocytes become more homogeneous from the point of view of immune information, which could indicate possible replicative senescence of some of these cells. Actually, it seems that lymphocytes from a woman of 115 years of age originated in a reduced number of stem cells, while the rest of stem cells lost their capabilities to differentiate (25).

If regarding replicative senescence it is not known whether it occurs *in vivo*, up-to-date data show that many cells could have undergone apoptosis in the course of aging, while other cells underwent meta-differentiations or alterations of cell functions (26), the aforementioned phenomena being involved in aging associated diseases (for example, the prostate adenoma. Changes in the organism's internal environment occuring in aging (hormonal, biochemical) seem to result in changes of cellular functionality.

In conclusion, replicative senescence would characterize cultivated cells by representing a likelihood of division under certain *in vitro* conditions, hence providing only orientative information on *in vivo* phenomena. Cellular senescence *in vivo* would have several functions as in wound recovery, embriogenesis but also age related pathology (27). Although (senescent) cells do not further divide, they are metabolically active, secrete pro-inflammatory cytokines, express specific gene and are positive for beta-galactosidase (28). In senescent cells there are also morphological changes, these cells being larger and more flattened. Clearance of senescent cells led, in an experimental model that used transgenic mice, to increase of defense against aging associated diseases (29).

The stationary phase model of aging in cells

The "Hayflick model" is very often used in a series of studies regarding cellular and molecular mechanisms of the aging process.

However, results were not always in accord with those obtained in studies concerning *in vivo* ageing. Even Hayflick supposed that cells *in vivo* never wear out their replication potential. In other words, the organism does not grow old because its cells reach the "Hayflick limit".

Since the start of our research on cell culture, we designed a system assuming that the stationary phase model of aging in cells may provide a series of relevant data regarding senescence (30,31).

Regarding cells of our system, because they did not replicate for a sufficient number of times as to reach the "Hayflick limit", replicative senescence was not of concern.

In our experiments, normal cells from cell lines or undifferentiated cells from primary cell cultures grew and divided until they formed a monolayer on the cell culture surface. They were maintained for up to 15-20 days and subsequently cells degenerated and died.

Thus, we distinguish phenotypes of "young" cells, as well as "old" cells respectively, which are distinct from the point of view of morphology, biochemistry and response to environmental factors.

Post-mitotic cells 6 to 8 days old are morphologically and physiologically clearly different from those aged 20 days old. During this complete time interval, cells are grown in a monolayer, without sub-cultivation, while the culture medium for cell growth is regularly replaced by the preservation medium (a medium without added serum). this way, age-induced changes can be detected by conducting longitudinal studies, in a relatively short amount of time.

Re-cultivation of cells "in confluence" restores the young cell phenotype

Subsequently, this type of study was mentioned by other researchers, who pointed out using similar experimental models, several age-related changes at cellular and molecular levels (32).

With regard to the aforementioned, there are some comments that should be made.

In vitro studies using postmitotic cells "in confluence:" have shown that these cells undergo multiple genomic changes during their lifespan, and the pathways of their response to environmental factors is similar with that of cells removed from old organisms. Our studies showed that cells being "in confluence" for a longer amount of time change their phenotype, "age", and become more sensitive to stressors (33,34).

On different note, human fibroblasts that underwent a higher number of divisions became more and more sensitive to various mechanisms which determine contact inhibition during cell growth. It is known that this inhibition appears in post-mitotic cells that are in confluence state, in cultures. It has been suggested that mechanisms involved in ceasing cell division occurring in the case of replicative senescence and those involved in contact inhibition would have a common final pathway (35).

Also, it is known that murine cells almost instantly undergo immortalization in cultures. Although these species are very sensitive (predisposed) to cancer, cell cultivation enhances the 'susceptibility to transformation of these cells even more (36).

Cultivation of cells seems, under these conditions, to represent an important stress for cells that determines alterations in the genome and either transforms cells into malignant ones or changes their phenotype after a certain amount of time. Cells in confluence are more exposed to environmental factors with a prooxidant effect. These are cultivated in atmospheres of 20% oxygen, whereas in the organism they are in a 2-5% oxygen atmosphere. Cultivation of human fibroblasts at oxygen physiological concentration increased their replication capacity by 20-30%, while in the case of rat fibroblasts, their cultivation in an atmosphere of 3% oxygen enhanced the number of divisions by 200-300%.

hese data, along with those showing a great variability of the replicative capacity of cells removed from various biopsies from the same individual, as well as the lack of correlations between the age of the donor and the number of divisions that cells can undergo in cultures (38), would naturally raise the question on what if the

Hayflick limit reflects the conditions of cell cultivation even to a greater extent than it has been previously considered,

The Hayflick limit, far from losing its importance, would result from the likelihood for a cell to continue to divide under certain conditions, the cell's genetic heritage coding for the susceptibility to pro-senescence factors, as well as significant past cell damages.

As shown above, simple cells cultivation, would determine an increase of cells' susceptibility to noxious factors. Although this marked susceptibility is different from *in vivo* aging, it could look like a much accelerated aging (38).

Under these conditions, we consider that morphological and physiological deterioration postmitotic cultured cells in confluence undergo with time could reflect in vivo changes of postmitotic or slow-dividing cells.

Why does this phenomenon occur in postmitotic cells in confluence? The answer to this question could be related to a stop to normal postmitotic cultured cells' growth. These cells divide until they form a monolayer on the bottom of a cell-culture plate. It has been assumed that compounds termed contact inhibitors are implicated. But there are cells in which this does not occur, cells which can form many layers in cultures. These are the cells transformed into malignant ones that have a defective cytoskeleton. Researches we conducted using HeLa cell line supported this finding.

On the other hand, an interesting experiment, which could explain the so-called "inteligent" reactivity (behavior) of cells in cultures showed that stimulating a cell with a solid object determines polymerization of actin in a given area, which leads to cells moving (39).

Our hypothesis in this case would be the following: cells set more points of contact with the bottom of the cell-culture plate than they would set with neighboring cells, which leads to a re-arrangement of their cytoskeleton different from that occuring *in vivo*. The new "architecture" of the cytoskeleton could be incompatible with vertical cell divisions. This state of the cytoskeleton could also influence the state of the genome and consequently, genes transcription. This phenomenon might explain some of the genome aberartions that occur in cultures, as well as the accelerated aging of these cells.

The involvement of some intercellular signals as inhibitors of contact is not excluded. The deformation of cells due to cultivation could have consequences on the secretory function of the cells.

Divisions at longer time intervals might increase the ratio of cells that undergo mutations incompatible with continuing replications. Indeed, in tissues with a high rate of cell divisions (the intestine), more mutations occur and accumulate with age than mutations in postmitotic cells (heart). But these regard only point mutations, whereas in the cardiac tissue re-arrangements of the genome accumulate (40).

These observations suggest that genetic material repair mechanisms are probably activated in mitosis and determine changes in the cytosol, which affect the state of the genome and by extention, gene transcription.

The above mentioned seems to be contradictory to the Hayflick hypothesis, according to which the attainment of the number of possible cell divisions that a cell can undergo may determine replicative senescence. But it appears that replicative senescence can occur due to the effects of environmental factors that damage the DNA (double strand-break), pro-oxidant agents, or through proto-oncogenes activation (41). Mitosis could repair damages in the genetic material before the extent of the damages reaches the critical threshold, which is incompatible with cell replication. Even though most somatic tissues from human adults completely lack telomerase, somatic tissues from adults,of other species (birds, mice, some vertebrates and marine invertebrates) synthesize considerable amounts of telomerase (42). At least in some of these cases telomerase activity could be associated with some repair mechanisms that could intervene at the DNA level during mitosis.

In conclusion, understanding the process of aging at cellular and molecular levels is key to development of therapeutic strategies related both to the physiological decline of aging and the prevention of chronic diseases associated to aging, the main purpose being that of a so-called "successful aging".

References

1. Hayflick L, Moorehead PS, The serial cultivation of human diploid cell strains, Exp. Cell Res, 1961;25: 585-621.
2. Hayflick L, The limited in vitro lifetime of human diploid cell strains, Exp. Cell. Res. 1965 Mar;37:614–636
3. Cristofalo VJ[1], Volker C[2], Francis MK[3], Tresini M[4], Age-dependent modifications of gene expression in human fibroblasts. Crit Rev Eukaryot Gene Expr.[5] 1998;8:43-80
4. Campisi J, From cells to organisms: can we learn about aging from cells in culture?, Exp Gerontol 2001;36: 607- 618
5. Miller JK, Escaping senescence: demographic data from the three-Toed box turtle (Terrapene carolina triunguis), Exp. Gerontology 2001;36: 829-832.
6. Campisi J, Replicative senescence: An old lives tale? Cell

1. http://www.ncbi.nlm.nih.gov/pubmed?term=Cristofalo%20VJ%5BAuthor%5D&cauthor=true&cauthor_uid=9673450
2. http://www.ncbi.nlm.nih.gov/pubmed?term=Volker%20C%5BAuthor%5D&cauthor=true&cauthor_uid=9673450
3. http://www.ncbi.nlm.nih.gov/pubmed?term=Francis%20MK%5BAuthor%5D&cauthor=true&cauthor_uid=9673450
4. http://www.ncbi.nlm.nih.gov/pubmed?term=Tresini%20M%5BAuthor%5D&cauthor=true&cauthor_uid=9673450
5. http://www.ncbi.nlm.nih.gov/pubmed/9673450

1996; 84: 497-500

7. Hayflick L, How and why we age, Exp Gerontol 1998; 33: 639-653.
8. Olovnikov AM, A theory of marginotomy. The incomplete copying of template margin in enzymic synthesis of polynucleotides and biological significance of the phenomenon. J Theor Biol 1973; 41 : 181–90
9. Greider CW[6], Blackburn EH[7], Identification of a specific telomere terminal transferase activity in *Tetrahymena* extracts. Cell.[8] 1985 ; 43: 405-13.
10. Shay JW, Bacchetti S , A survey of telomerase activity in human cancer, Eur. J. Cancer 1997; 33: 787–91.
11. Blasco MA, Telomeres in cancer therapy, J.Biomed.Biotechnol 2001;1:3-4
12. Bodnar AG, Oulette M, Frolkis M, Holt SE, Chiu CP, et al., Extension of life by introduction of telomerase into normal human cells, Science 1998 ; 279: 349-352.
13. Blasco MA, Mammalian telomeres and telomerase:way they matter for cancer aging, Eur.J.Cell Biol., 2003; 82:441-446
14. Bryan TM, Englezou A, Gupta J, Bacchetti S, Reddel RR, Telomere elongation in immortal human cells without detectable telomerase activity EMBO J. 1995; 14 : 4240–8
15. Henson JD, Neumann AA, Yeager TR, Reddel RR, Alternative lengthening of telomeres in mammalian cells, Oncogene 2002; 21 : 598–610
16. Pernaute B, Spruce T; Smith K M, Sánchez-Nieto J M; Manzanares M; Cobb B, MicroRNA control the apoptotic

6. http://www.ncbi.nlm.nih.gov/pubmed?term=Greider%20CW%5BAuthor%5D&cauthor=true&cauthor_uid=3907856

7. http://www.ncbi.nlm.nih.gov/pubmed?term=Blackburn%20EH%5BAuthor%5D&cauthor=true&cauthor_uid=3907856

8. http://www.ncbi.nlm.nih.gov/pubmed/3907856

threshold in prime pluripotent stem cells through regulation of BIM, Genes Dev 2014 28: 1873-1878

17. Jaskelioff M, Muller FL, Paik J-H, Thomas E, Jiang S, et al Telomerase reactivation reverses tissue degeneration in aged telomerase-deficient mice, Nature 2011; 469:,102–106
18. Gomes NM, Ryder OA, Houck ML, Charter SJ, Walker W, et al,. Comparative biology of mammalian telomeres: hypotheses on ancestral states and the roles of telomeres in longevity determination, Aging Cell 2011;10:761-768. doi: 10.1111/j.1474-9726.2011.00718.
19. Cherif H, Tarry JL, Ozanne SE, Hales CN, Ageing and telomeres: a study into organ- and gender-specific telomere shortening, Nucleic Acids Res 2003; 31 : 1576–1583.
20. Renault V, ThorneL-E, Eriksson P-O, Butler-Browne G and Moul V, Regenerative potential of human skeletal muscle during aging, Aging Cell 2002; 1 : 132–9.
21. Blasco MA , Telomeres and human disease: ageing, cancer and beyond, Nature Reviews Genetics 2005; 6 : 611–22.
22. Crews DE and Stewart JA, Human longevity and Senescence, in Muehlenbein M P (editor) Human evolutionary biology, Cambridge University.
23. Diaconeasa AG, Rachita M, Stefan van Staden R-I, The new hypothesis of aging, Medical Hypothesis, 2015 in press.
24. Hayflick L, How and why we age, Exp. Gerontol 1998; 33: 639-653.
25. Holstege H, Pfeiffer W, Sie D, Hulsman M, Nicholas T J., Lee CC et al, Somatic mutations found in the healthy blood compartment of a 115-yr-old woman demonstrate oligoclonal hematopoiesis, Genome Res 2014; 24: 733-742; Published in Advance April 23, 2014, doi:10.1101/gr.162131.113
26. Macieira-Coelho A, Relevance of in vitro studies for aging of

the organism. A review.[9] Gerontol Geriatr. 2001; 34 : 429-36
27. Burton, D G A, Krizhanovsky V , Physiological and pathological consequences of cellular senescence, Cellular and Molecular Life Sciences 2014; 71: 4373–4386.
28. Rao MS[10], Mattson MP[11], Stem cells and aging: expanding the possibilities Mech Ageing Dev[12] 2001 ; 122 : 713-34.
29. Campisi, J, Aging, Cellular Senescence, and Cancer, Annual Review of Physiology 2013; 75: 685–705.
30. Ionescu T, Rachita M.,Tudor M, Andrei V, Activitatea enzimatica a fosfatazei acide in culture de celule renale tratate cu Gerovital H3,Proceedings of Vth European Symposium on Basic Research in Gerontology,Weimar ,1976,p 213-218.
31. Aslan A. Constantinescu E.,Rachita M., Andrei V,Ionescu T, Lactate dehydrogenase isoenzymes from monkey renal cell culture,action of Gerovital H3 treatement,Rom. J.of Gerontology and Geriatrics,1982,4,tome 3.
32. Khokhlov AN. Stationary cell cultures as a tool for gerontological studies. Ann N Y Acad Sci 1992; 663: 475-6.
33. Hanu N, Rachita M,Carasava M, Research concerning the stress and aging in systems of cells cultures in the presence of activated waters by physical fields using an original method,14 thInternational congress on Cybernetics,Namur (Belgium)21-25 august, 1995 ,Proceedings.
34. Rachita M,Hanu N,Rusu C,Gradinaru D,Studies concerning the effects of free radical generation on cultured cells and the cytoprotective effect of Gerovital H3, Rom.J.of Gerontology

9. http://www.ncbi.nlm.nih.gov/pubmed/11828879

10. http://www.ncbi.nlm.nih.gov/pubmed/?term=Rao%20MS%5BAuthor%5D&cauthor=true&cauthor_uid=11322994

11. http://www.ncbi.nlm.nih.gov/pubmed/?term=Mattson%20MP%5BAuthor%5D&cauthor=true&cauthor_uid=11322994

12. http://www.ncbi.nlm.nih.gov/pubmed/11322994

and Geriatrics, 2000,vol.22,1,p.22-31.

35. Macieira –Coelho A, Contributions made by the studies in vitro for understanding of the mechanisms of aging.,Exp.Gerontology,1993,28,1-16.
36. Macieira –Coelho A, Cancer and aging at the a cellular level.In Cancer and aging, Macieira –Coelho A and Nordenskjoeld B.(editors,pp 11-37,CRC Press Inc,Boca Raton ,FL,1990.
37. Campisi J, Kim SH, Lim CS,Rubio M, Cellular senescence, cancer and aging:The telomere conection,Exp.Gerontol.2001,36:1619-1637.
38. Cristofalo V.J. Cell culture aging:Insights for cell aging in vivo?Aging.Clin.Exp.Res.,1999,11,1.
39. Banes AJ, Tsuzaki M, Yamammoto J, Fisher T, Mechanoreception at cellular level:The detection,interpretation and diversity of responses to mechanical signals, Biochemistry and Cell Biology,1995,vol 73,nr.7-8.
40. Kirkwood TBL,Proctor CJ, Somatic mutations and aging in silico, Mechanisms of aging and development,2003,124,85-92.
41. Campisi J, Cellular senescence as a tumor suppressor mechanism,Trends in Cell Biology,2001,11,27-31.
42. Campisi J, Cancer,aging and cellular senescence,In vivo, 2000, 14, 183-188

CANCER AND AGING

The fact that there is a link between cancer and aging is well known. Increased lifespan is the main factor in increasing the incidence of cancer, which has intensified in recent decades (1). According to some data, after the age of 45, every 9 years, the incidence of cancer doubles. There is a somewhat causal link between cancer and aging, probably both being consequences of the same phenomenon, but there are also differences related to mechanisms and first of all, clinical data. We will briefly discuss these aspects, the similarities and differences between the two degenerative conditions.

Aging and the replicative limit of cells

One of the characteristics of malignant cells is immortality. They divide indefinitely in cultures, the HeLa cell line has been providing large amounts of cells for decades for research. Normal cells have a finite number of divisions in cultures, depending on the species to which the cultured cells belong. In general, cells from long-lived species are able to divide more times than those from less long-lived species. For example rat cells can divide less than human cells, which in turn show fewer divisions than those of the turtle (2). However, these data represent trends, because differences were found in the ability to divide between cells from the same individual, taken from different tissues. This maximum capacity of cells to divide in cultures is called the replicative limit or the Hayflick's limit, after the name of the researcher who discovered it in 1961. Until then, it was considered that all cells could divide indefinitely in cultures, according to the results of experiments in nineteenth century, performed by Francis Carell. In these experiments, Carell brought new cells with the culture medium, which created this confusion (2).

Leonard Hayflick, starting from his discovery, later launched two hypotheses: the first is related to the involvement in aging of the replicative limit. In other words, aging would result from this cellular replicative limit, more precisely from their inability to divide at some point. The attainment of that state is called cellular senescence. Here again things must be treated with caution because, at least in the human species, cellular senescence would be reached at ages beyond the attested maximum longevity (3, 4).

Hayflick's second hypothesis concerns the possible evolutionary cause of the replicative limit. He assumes that this appeared as a protection against cancer. Once the cells have accumulated many errors, senescence, cessation of division would function as a kind of natural anti-cancer mechanism.

What is the mechanism of cellular senescence and the link between it and cancer? An important link between senescence and cancer would be telomerase.

At each division, the (non-informational) ends of the chromosomes, called telomeres, shorten. They contain repetitive DNA sequences, the role of which seems to be to protect telomeres from fusion. Telomere shortening at each division occurs due to DNA polymerase that binds to DNA strands, and the site of attachment cannot be replicated for this reason. The solution against telomere shortening is an enzyme called telomerase, which repairs telomeres. Telomerase is secreted in large quantities by embryonic stem cells, but also by rapidly dividing adult tissue cells, such as skin cells, mucous membranes, hematogenous marrow, etc. (2). Interestingly, many malignant cell lines (90%) secrete telomerase, giving them immortality (5). In cultures, cell transformation occurs near the point of senescence (6). For example, in the case of double-stranded ruptures, the repair takes place by realigning the chromosomes, which fuse, then recombinations take place and, as a result, genetic and / or epigenetic mutations occur that lead to the activation of genes such as telomerase. Is telomerase necessary and sufficient for malignancy? Elisabeth Blackburn, one of the researchers who discovered telomerase, a Nobel Prize winner in 2009, said the telomerase gene is responsible for regulating the activity of another 70 genes important for cellular function, which would play important roles in malignancy. She also showed that the integrity of the RNA contained in telomerase is essential for most of the harmful effects of cancer related to telomerase, even if those cells have an alternative mechanism for maintaining telomerase. This aspect recommends telomerase as a therapeutic target.

Telomerase saves malignant cells from senescence, but can we assume that it is directly involved in the appearance of the malignant phenotype? Attempts are now being made to develop telomerase vaccines in the treatment of cancer (7). The problem would be that such a vaccine would severely affect rapidly dividing cells. Moreover, recent data show that the efficacy of vaccines against malignant cells is limited by their very metabolism. They secrete metabolites that affect the activity of inducible regulatory T lymphocytes (Tregs), which block the immune response to tumor antigens (8). Also, the problem becomes even more difficult with aging, when changes in the immune system make the action of vaccines even less effective (8).

An even more important aspect is that some cell lines have an alternative mechanism for telemore repair, which is alternative to activating telomerase, called ALT (Alternative Lengthening of Telomeres) (9). In their case, the telomerase vaccine would be totally ineffective. The question would be whether there is only one or more such alternative mechanisms.

If the relationship between telomerase and cancer is not very direct, the same can be said about the relationship between telomerase and aging. Rats, short-lived animals, have very long telomeres (10). Probably these occurred precisely because of the increased risk of recombination caused by the fusion of chromosomes, more precisely the instability of the genome of these species.

Moreover, cell senescence can be achieved in some murines in the absence of telomere shortening and even in the presence of large amounts of telomerase (11), but there are also human cell types that either enter senescence despite the presence of telomerase or have telomeres so long at senescence that telomere shortening does not seem to play any role (12).

Cancer and evolution

Hayflick hypothesizes that cellular senescence has evolved to protect against cancer. The natural question would be: what is more harmful to the body, aging or cancer? In mammals at least, cancer kills faster than aging, but aging certainly kills. Would it have been advantageous for evolution that, due to the risk with a certain probability of developing cancer,to make the cells reach senescence and the body to reach aging? When does malignancy occur in evolution? In Coelenterates (hydra), animals with extraordinary regenerative power, no malignancy has been reported, but in insects (*Drosophila*) there are tumors. Hydra severely irradiated so that the cells lose their ability to divide, if part of it is removed, it still retains its ability to regenerate, so the remaining cells are polarized giving rise to a smaller hydra, but with a normal shape (13). Hydra is composed almost entirely from stem cells (14). On the other hand, adult *Drosophila* is postmitotic.

Also, in reptiles, long-lived animals (e.g. Galapagos tortoise) and/or with a high regenerative power (eg *Axolotus mexicanus* or water salamander), although the regeneration mechanism is different from that of Coelenterates, a high resistance to pollutants has been reported. Those pollutants do not increase the risk of cancer as much as in mammals. Moreover, birds, a class of long-lived animals compared to mammals at the same body mass, although they have a higher metabolic rate, are also more resistant to cancer. Their cells in cultures, treated with mutagenic/carcinogenic factors, go through several stages of malignancy, after which, most of the time, they stop, failing in their attempt to transform (15). On the other hand, laboratory rodent cells spontaneously transform (become malignant) in cultures at a very high rate, about 106 times higher than in the case of human cells, for example (16).

What would be the connection between the risk of malignancy and the appearance of cellular senescence in evolution? The above data show that animal species that are long-lived or have a high regenerative capacity have a lower risk of malignancy than those with a reduced longevity. Should the latter have early cellular senescence, so faster aging and shorter life consequences of the risk of cancer? Or are they consequences of the same phenomenon, for example genome instability?

If we take into account the first variant of response, we consider that, at some point in evolution, in some species at high risk of malignancy, cellular senescence would have appeared through a mutation. What were the chances of this being imposed in evolution, given that it is associated with a decrease in fitness? One of the criteria for establishing the phenomenon of aging in species that do not show detectable characteristics of this process is the decrease in fertility associated with the age, along with increased mortality (17). Aging is a decrease in fitness, so it could be imposed in evolution only if malignancy was a strong factor of negative selection, i.e. causes a drastic decrease in fitness in the population. Under these conditions, the phenomenon of cellular senescence offered an advantage over a period of time, in the sense that individuals carrying this mutation could reproduce for a period of time before their reproduction ceased or diminished greatly.

In order to establish an answer to this question, a precise mathematical model should be created. But this model would start from the premise that in a stage of evolution, cancer was a factor of decline in fitness so strong that the appearance of cellular senescence, with its consequence, aging of the body, become a selective advantage. That means that the individuals who grew older should have more offspring than those who developed tumors.

The current data do not suggest that malignancy has ever evolved into a strong negative selection factor. Most likely, it is not a selection factor in any of the current species, because even in the most affected, such as rats, this disease occurs in old age, when it is no longer an important subject to natural selection.

According to the hypothesis of antagonistic pleiotropy (2), in old age, many genes would be expressed which in youth would be subject to natural selection. The senescent phenotype would result from the expression of these genes which were harmful to the body. Those genes would have escape the action of natural selection after the body no longer reproduced. But if we consider Finch and Austad's criterion of senescence, according to which aging should be associated with a reproductive decline, then we arrive at a circular reasoning. If aging were not associated with reproductive decline, those genes would no longer be expressed, so aging would not occur.

As in evolution, but also in ontogeny (aging in this case), the risk of malignancy seems to be associated with a reduced capacity for regeneration or maintenance (increased in long-lived species) it seems very likely that aging and cancer would be consequences of the same phenomenon, and a certain pattern of aging would be associated with an increased risk of malignancy.

We advance the hypothesis of rapid aging as related to a possible correlation between increased risk of cancer and senescence. A certain pattern of aging would actually result from certain ontogenetic peculiarities. We can hypothesize that the existence of stages in the development of the organism, in which there is rapid cell proliferation, with superficial corrections of genetic defects, would cause epigenetic changes and alteration in cell physiology able to determine later in life the senescent phenotype.

Aging would be the consequence of a functional pleiotropy. Pleiotropy refers to the multiple actions of a gene. In this case, the same type of cellular response, useful at some point in the development of the body, would cause harmful phenomena at another stage of life. But this response would involve the concerted action of several genes (with a high degree of ontogenetic conservation) in response to environmental factors with adaptive importance. We can assume that, if during development there are processes of rapid proliferation without differentiation, later in life the chances of rapid aging with an increased risk of malignancies will increase. A more detailed meta-analysis, as well as the experimental verification, could test this hypothesis.

In support of this are data, related to the very large difference in the incidence of cancer in humans (*Homo sapiens*), compared to the chimpanzee (*Pan troglodytes*) in the laboratory (18), i.e. 20% in the populations of developed countries vs. 2%. This difference would be explained according to the hypothesis presented above, by a certain pattern of human development, which comprises stages of intense cell proliferation with reduced differentiation. Heterochronism (alteration of development time) would result from these phenomena. The human baby at birth, although it is 17 months immature compared to the chimpanzee baby, has a weight almost 2 times higher, with a very high development of some tissues (brain, adipose tissue). This particularity of development, which would result in the appearance of defining characters for the human species, would come from the overactivation of a biochemical pathway of insulin / IGF1 signaling, involved in cell proliferation and survival (18). There is a positive correlation between IGF1 levels in adulthood and cancer incidence (19, cited by 18).

In humans, aging has an accelerated pace compared to the general pace of development, in addition in humans degenerative diseases that do not exist or are rare in chimpanzees often occur during aging (type II diabetes, Alzheimer's disease, autoimmune diseases, etc.). In all these diseases, the above-mentioned metabolic pathway is altered, as in other typical human diseases (mental illness, autism). Key enzymes in this pathway are constitutively activated in cancer. The easy activation of this metabolic pathway, useful in development, when it appears later in life, could determine the mentioned pathological phenomena.

Cellular stress in aging and cancer

Transformed cells have a certain characteristic appearance in cultures, in the sense that they do not grow in a single layer, like normal ones, they create three-dimensional local agglomerations. These cells do not present the so-called contact inhibition, characteristic of normal cells. Malignant cells show a temporal-spatial alteration of the normal way of growth, caused by changes in gene expression, with consequences on protein synthesis. Their cytoskeleton is altered, among other factors. Also, their response to environmental factors, which makes them dysfunctional and at the same time dangerous, is altered. Experiments in our laboratory since the 1990s with HeLa cells show they are refractory to cellular stress.

With the help of an original method (20), the changes in cell diameter under the influence of chemical factors (drugs, etc.), physical (temperature, pH, etc.) were highlighted. The principle of this method is that the change in cell diameter is measured by the variation of the optical density of the cell monolayer, recorded using contraction curves, called "optograms". Normal cells, in the presence of an environmental factor such as the above, change their cell diameter, I.e. "contract". Experiments performed in the presence of numerous stressors show that the cellular response to stress changes depending on the age of the cultures, becoming weaker and weaker starting with the cells "young" (3-4 days), "mature" (10-12 days), to the old ones (over 2 weeks). In these experiments, the cells are postmitotic (quiescent), not senescent in the sense of Hayflick's limit (see above).

Malignant cells (HeLa) do not respond to the cellular stress factors we studied. The main reason is probably the alteration of their cytoskeleton, because the disorganization (reversible depending on the duration of action of the stressor) of the cytoskeleton is the cause of cell contraction (21). Although, other important factors, involved in cell communication (some Calcium channels) may be altered, as our results suggest (22). But this is not the only altered cellular function in malignant cells, which synthesize specific proteins, abnormal to the tissues to which they belong, some present during embryonic development (eg alpha fetoprotein).

In malignancy, epigenetic changes occur in the affected tissues, which precede the appearance of tumors (23). Probably these epigenetic changes are then responsible for recombination, chromosomal aberrations, gene mutations, by changing the architecture of the genetic material. Epigenetic changes also occur in aging, including a reduction in gene methylation (24). Aging would be a kind of reduction of cellular specialization (differentiation), the defining functions of specialized cells being reduced.

By rapid proliferation, with poor differentiation, malignant cells act similarly to embryonic cells or cells involved in regeneration in vertebrates. The pattern of tissue regeneration in vertebrates is represented by dedifferentiation, followed by cell proliferation and redifferentiation (14). In fact, cancer is a cellular dedifferentiation, followed by proliferation, but without the necessary redifferentiation of normal tissue.

An interesting link between longevity, Hayflick limit, telomerase, aging and cancer is suggested by the distribution of telomerase in such species. Although the replicative potential of the cells from more turtle species varies, telomerase was not found in any other tissue except for the gonads in any of these species, except for two ornate box turtle

(*Terrapene ornata*) hatchlings had telomerase activity in two organs other than the gonads (25). But telomerase was found in all of the tissues (including brain) of a teiid lizard. The presence of telomerase in somatic tissues of an adult terrestrial vertebrate with no tendency for non-viral cancer is interested, as authors pointed out (25). Teiid lizards are often parthenogenetic, and, as other parthenogenetic reptiles, This suggest that telomerase may play a part in DNA and tissue repair and regeneration.

Cancer can be said to be a failed attempt at regeneration in organisms that do not have a very high capacity in this regard. It can be seen, as a bye-product of evolution.

In conclusion, we can say that between cancer and aging there are many similarities of mechanism, most likely being consequences of the same phenomenon. The theoretical understanding of these conditions and their evolution will lead to the discovery of therapeutic or prevention strategies.

References

1. Anisimov VN, Biology of Aging and Cancer. Cancer Control 2007, 14(1) 23-31.
2. Campisi J., From cells to organisms: can we learn about aging from cells in culture? Experimental Gerontology 2001; vol. 36 (4-6) pp. 607-618.
3. Mondello C, Petropoulou C, Monti D, Gonos ES., Francheschi C., Nuzzo F ,Telomere length in fibroblasts and blood cells from healthy centenarians, Exp. Cell.Res.,1999; 248, 234-242.
4. Mikhelson VM. Replicative mosaicism might explain the seeming contradictions in the telomere theory of aging, Mechanism of Ageing and Development 2001; 122 ,1361-1365.

5. Bryan TM & colab.,The telomere legthening mechanism in telomerase-negative immortal human cells does not involve the telomerase RNA subunit. Hum. Mol. Genet.1997; 6 (6) 921-926.
6. Klapper w., Parwarwsh R., Krupp G., Telomere biology in human aging and aging syndromes, Mechanisms of aging and development 2001;122, 695-71.
7. Vonderheide, R ., Prospects and challenges of building a cancer vaccine targeting telomerase, Biochemie, 2008; 90(1), 173-180
8. Gravekamp C, Kim SH., Castro F., Cancer vaccination: Manipulation of immune responses at old age, Mechanism of aging and development 2008;
9. Bryan TM & al., Evidence for an alternative mechanism for maintaining telomere length in human tumors and tumor-derived cell lines. Nat. Med.1997; 3 (11), 1271-1274.
10. Serrano M, Lin AW., McCurrach ME., Beach D ,Lowe SW, Cell 1977; 88, 593-602.
11. Carman CA., Afshari CA.,Barett JC., Cellular senescence in telomerase expressing Syrian hamster embryo cells,. Exp. Cell. Res.1977; 244, 33-42.
12. Kiyono, T., Foster, S.A., Koop, J.I., McDougall, J.K., Galloway, D.A. and Klingelhutz, A.J., 1998. Both Rb/p16INK4a inactivation and telomerase activity are required to immortalize human epithelial cells. *Nature*, *396*(6706), pp.84-88.
13. Hicklin J, Wolpert L. Positional information and pattern regulation in hydra: the effect of y-radiation. Embryol Exp Morph 1973;30:741–52.
14. Stocum DL. Tissue restoration through regenerative biology and medicine. Adv Anat Embryol Cell Biol. 2004;176:III-VIII, 1-101, back cover. doi: 10.1007/978-3-642-18928-9.

PMID: 15079897.

15. Macieira-Coelho A. Contributions made by the studies in vitro for understanding of the mechansims of aging. Exp. Gerontology 1993; 28, 1-16.\
16. McCormick JJ, Maher VM. Malignant transformation of mammalian cells in culture, including human cells. Environmental and molecular mutagenesis. 1989;14(S16):105-13.
17. Finch CE., Austad SN.,History and Prospects: symposium on organisms with slow aging, Experimental Gerontology 2001; vol. 36, pp..593-597.
18. Diaconeasa AG., Civilizatia foametei/O alta abordare a umanizarii, Sitech, 2007; pp. 56, 60.
19. Elias SG. et al., The 1944-1945 Dutch Famine and susequent Overall Cancer Incidence, Cancer Epidemiology Biomarkers& Prevention 2005; vol. 14, pp. 1981-1985, citat de Diaconeasa A.G, Civilizatia foametei/O alta abordare a umanizarii, Sitech, 2007, pp. 64, 114.
20. Hanu N.,Rachita M.,Cell culture procedure and the optometric method to study cells "in vitro",Rom.Archives of Microbiology and Immunology,1999;58,147-155.
21. Răchită M, Cercetări privind procesul de îmbatrânire pe sisteme celulare in vitro, Teză de doctorat, 2005; Universitatea din Bucureşti, Facultatea de Biologie
22. Diaconeasa G , Răchită M. Cercetări privind mecanismele stresului celular intrun model experimental original. Anal Univ Bucuresti, 2003.
23. Herceg Z., Epigenetics and cancer: towards an evaluation of the impact of environmental and dietary factors, Mutagenesis 2007; 22 (2):91-103;
24. Wilson VL, Jones PA., DNA methylation decreases in aging but not in immortal cells, Science, 1983: Vol. 220. no. 4601,

pp. 1055 – 1057.

25. Christiansen JL, Henderson ER, Budke B, Lynch M, Lu Q, Johnson J. A final report of studies of the Hayflick limit in Reptiles, a test of potential immortality. Proceedings of the Iowa space grant consortium. 2001;10.

PERSPECTIVES REGARDING DEGENERATIVE DISEASES TREATMENTS AND AGING

AS PROVIDED BY THE STUDY OF STEM CELLS

Introduction

Stem cells almost like no other discovery in biology and medicine have raised great interest, controversy and concern about ethics in the scientific and non-scientific world. What are stem cells?

The stem cell is defined as an undifferentiated cell, which through mitotic division results in two cells having the capacity to either remain in the stage of stem cells (thus preserving the undifferentiated character) or to differentiate following successive divisions. In this way, the cell can divide symmetrically, in which case two new cells or two differentiated cells result, or it can divide asymmetrically and a stem cell and a differentiated cell result. The capacity of the stem cell to produce another identical cell has been termed *self-renewal,* whereas the capacity to result in two different types of cells is termed differentiation. Sperm cells and ovocytes are stem cells.

Researches on stem cells developed extensively in the last years because there has been great hope for treatment of some degenerative diseases using stem cells, replacement of tissues damaged in the course of diseases or by factors causing damage and not in the least, for treatments in aging [1].

Research data have suggested that stem cells, when introduced in an adequate cell medium, under the influence of growth factors can transform themselves in a desired type of tissue. When stem cells are introduced in the organism, they can reach to the injured areas where they transform themselves in cells typical of that tissue because of the tissue generated microenvironment, thus improving the function of the tissue. For instance, by use of this procedure, there is hope that Parkinson's, Alzheimer's diseases, diabetes, cancer, heart failure will be (better) treated, to the extent that knowledge in the field advances [2].

There were identified cellular signals that control proliferation, differentiation and survival of stem cells (growth factors, cytokines, adhesion molecules and intracellular mechanisms regulating stem cells (cell signal transduction secondary messengers, transcription factors, telomerase) [3].

New techniques of genetics pointed out a series of genes whose expression has been related to the particular stem cells' state. Genes expressed by different types of stem, embryonic, hematopoiesis stem cells are diverse and their overlapping is very little and hence, a great complexity of these genes' functions.

Research endeavors have aimed at discovering and understanding genes' functioning that maintain stem cells undifferentiated, as undifferentiating is considered the key to the stem cells' function of repair.

The more a stem cell is undifferentiated, the more it produces many types of cells through transdifferentiation. From this point of view embryonic totipotent stem cells are distinguished from adult stem cells of various organs and tissues, almost all (bone marrow, adipose tissue, muscle, skin), which are multi-potent. The aforementioned preserve their distinctiveness as stem cells (due to the modality of division), but have passed through a few stages of differentiation.

Which are the factors involved in maintenance of stem cells' toti-potency? The Nanog gene is noteworthy as it appears to be responsible for preserving an undifferentiated state of embryonic stem cells. This gene codes a transcription factor the overexpression of which is involved in preservation of the multi-potency state of human embryonic stem cells, whereas this gene's absence leads to differentiation. Other such transcription factors are oct-4 si SOX2, the expression of which together with that of Nanog mutually potentiate thus blocking differentiation pathways of the embryonic stem cells [4].

Certainly, embryonic stem cells are preferred to adult stem cells from various organs. But due to ethical considerations, in some countries, studies on embryonic stem cells have been forbidden partially or totally. Under these circumstances, there are attempts at developing treatment methods based on adult stem cells. *In vitro* studies seem promising, but nevertheless some problems still have be solved.

Problems raised by the treatment with stem cells

Firstly, there is little knowledge about stem cells functioning *in vivo*. One study generated many controversies because it presumed that in fact transdifferentiation of stem cells which are introduced in the organism would take place in target organs through fusion of undifferentiated and differentiated cells of that organ tissue. The result of these processes would be polyploid nuclei along with all risks of genetic instability, which can lead to cancer.

Under these circumstances, the problem of safety of an eventual treatment with stem cells has been raised. Other studies declined these results and reported that transdifferentiation without cellular fusion was obtained [5]. Anyway, it is not clear how transdifferentiation takes place and moreover, the *in vivo* correspondent of embryonic stem cells is not known yet. The aforementioned cells obtained *in vitro* derived from blastocyst, which permanently preserves its state of pluripotency.

A major problem regarding transplantation of embryonic or pluripotent stem cells, including those with an induced pluri-potency, is the occurrence of benign tumors, teratoma, which can turn into malignant tumors, teratocarcinoma [6]. In theory, at present, data converge towards the idea that from every human cell, any other type of cell can be produced. The aforementioned seems to be a normal fact considering that life relates to chemistry and with appropriate reagents, if one has the knowledge about how to add these reagents to a medium and control the later mentioned, then they can obtain the desired reaction products. The aspiration (dream) to produce multipotent stem cell lines already became a fact. Two research teams, Japanese and American, produced multipotent stem cell lines from differentiated epithelial cells using transcription factors Oct3/4, Sox2, c-Myc, and Klf4 [7] and Oct4, Sox2, Nanog and Lin28, respectively [8].

The problem of stem cell transplantation regards differentiation. It is difficult to produce the desired type of differentiated cell because cells' differentiation is not uniform. Sometimes cells appear to differentiate to totally undesired directions [9].

Malignization itself regards differentiation. Transcription factors involved in maintaining undifferentiation, pluri-potency of embryonic stem cells have been encountered in tumors oct-4 [10], Nanog [11.12], Sox-2 [13], thus being useful in defining such tumors.

Why does this phenomenon occur? The biochemical hypothesis we advanced [14] could explain it in that aging would result from the loss or decrease of certain cellular signals that are related to particular functioning of an organism, also having evolutionary and ecological implications. In the course of aging, signals involved in cell differentiation and repair, which are stimulated under certain conditions of cellular functioning, diminish at a rate that differs among various species Maintenance of a young phenotype would result from stimulating these reactions through organism's functioning or by slowly diminishing them.

Data regarding maintenance of pluripotent stem cell lines are in accord with the above mentioned hypothesis. There are two theories concerning maintenance and loss of totipotent stem cell undifferentiation. According to one of them, the type of division plays the main role in maintaining and loss of undifferentiation. Stem cells divide symmetrically and same stem cells result but they also divide asymmetrically and a stem cell as well as a progenitor cell result which after several divisions differentiates into a mature cell. In this sense, the polarizing cell, plasma membrane and cytoplasm polarization would play roles in distributing membrane proteins (receptors) [15] and gradient of cytoplasmatic substances. The role of polarizing in (cell) regeneration is well-known and as related to this, induction of polarization could lead to restoring stem cells' capacity to differentiate.

According to the biochemical hypothesis of aging, it is polarization, among other factors, which is lost once a number of divisions take place and this fact would explain progenitor cells' but also stem cells' diminished capacities to differentiate with age.

Another theory is related to biochemical niche of stem cells in the organism. When these cells are removed from that niche, they lose their capacity to differentiate. Experiments using Drosophila have been a good illustration of this view [16, 17], which is also according with the advanced biochemical hypothesis on aging. During the stage of aging, actually any cellular microenvironment, including that of stem cells is changing biochemically through diminishment of signals. Loss of capacity to differentiate occurs with age, probably due to lack of environmental stimuli and division which is less asymmetrical. Stem cells differentiate and are depleted because environmental factors change. Division itself abates signals and this fact is the reason for asymmetry (actually cell polarization) being important. Asymmetry is reduced through division, if it is no longer stimulated.

Prediction would be that adequate signals could restore stem cells capacity to differentiate. In this sense, a study of embryology would be very useful. It is not enough stem cells be very undifferentiated, pluri or totipotent as embryonic stem cells are, what counts is how broad their differentiating capacity is in terms of cell polarization and capacity to specifically synthesize (substances).

Also, taking into account similarities of stem cells to malignant cells, in presence of adequate factors not only cancer could be treated, but malignant cells with their indefinite capacity to divide, after proliferation, could turn into stem cells and this fact provides a huge capacity to regenerate.

It is important that in adult organisms, stem cells are being found permanently. In the brain of old people there are neuronal stem cells. Therefore, adult organisms use stem cells for all time in order to repair. Then the ideal solving would be *ad hoc* producing stem cells at the site where stem cells are needed or at least stimulation of existing stem cells' function.

The question remains whether these cells have to be very undifferentiated. It may be possible that stock of stem cells from certain organs be depleted, meaning the number of stem cells decreases until their differentiation becomes impossible. Probably this phenomenon takes place in some degenerative diseases and menopause [18] and for instance, apoptosis under the influence of auto-immune factors would be causing it. Even in this case, somatic stem cells undergo biochemical changes through deficit of signals, which makes them vulnerable to apoptosis, their differentiation becomes impaired and even these cells differentiate, their function is impaired.

In other words, therapy with stem cells could aim at restoring stem cells potential to differentiate by use of biochemical treatments with transcription factors and eventually substances that stimulate synthesis of transcription factors, which finally lead to expression of genes responsible for restoring the initial functionality of stem cells. When the number of stem cells is too reduced, local proliferation of stem cells without their differentiation, could be stimulated. Under these conditions, restoring stem cell functioning could be the simple and efficient solution.

This idea, which actually is not circulated, copes with inconveniences of treatments with stem cells by introducing these cells in the organism. Apart from risks, treatments with stem cells implying the aforementioned and other unknown risks, the very high cost of such a highly personalized treatment has to be taken into consideration.

Perspectives regarding cell therapies in neurodegenerative diseases

Even at present levels of knowledge, stem cells are being used to treat successfully some diseases. Ombilical stem cells removed at birth are utilized in treating some forms of leukemia and in some countries, there are such stem cell biobanks. Cells of this type are relatively cheap as they are present at birth, and there is no need for sophisticated procedures to remove them, nor do these aforementioned raise ethical problems.

Cell therapies utilizing stem cells for neurodegenerative diseases is at an early stage [19]. The major challenge in using them is that of identification of molecular determinants of stem cell proliferation and control exertion on these stem cells and their untoward genetic alterations.For neurons from stem cells it is necessary to induce their functional integration in existing afferent and efferent synaptic networks.

Brain's potential for self-regeneration is unexplored. New technologies of genetically labeling of stem cells are needed for precisely establishing where neurogenesis takes place and what types of cells are involved.

Genomic and proteomic techniques will aim at identification of molecular markers which regulate different stages of neurogenesis through a concerted action. It is also necessary to develop models utilizing experimental animals and mimicking human diseases and only based on this knowledge, strategies for therapies to use in humans could be elaborated. Not in the least evaluation of the ratio risk/benefit is needed before utilizing the stem cell therapy in humans[20].

In rodents, but also in humans some data suggested that stem cells could be stimulated to produce new neurons, a fact which is contrary to the canon of neurons that do not regenerate [1]. Dr. Kunlin Jin has shown that neuronal stem cells can be handled by use of growth factors such as FGF-2, EGF, VEGF and SCF even through nasal administration. Also, Dr. Jin and coworkers have demonstrated that cerebral ischemia (focal cerebral ischemia) was reversed in old mice by adminstration of plasma from young mice [1]. These results were as well according to the biochemical hypothesis of aging, whereas parabiosis (old animals were given blood from young animals) was very important in designing the hypothesis of aging [14].

If we were to summarize, the major succes of stem cell treatment would be the treatment of multiple sclerosis by transplant of bone marrow stem cells [21]. The aforementioned regards autologous grafts of bone marrow stem cells handled using a procedure which was proposed twenty years ago, but in the meantime developed through an increasingly efficient isolation of stem cells. After removing bone marrow stem cells, a patient's immune system is inactivated by the chemotherapy, similarly to what happens during treatment of leukemia and subsequently stem cells are transplanted. Conditions of many patients who were selected by tight criteria, greatly improved when this transplant treatment was performed under the frame of a randomized multicenter study.

This treatment is certainly very invasive and not recommended for patients aged over fifty years because it affects other organs. A great disadvantage of treatments with stem cells is the requirement for immunosupression in order to remove the patient's cells or so that his immune system does not attack the transplanted stem cells. The patient's own stem cells are a solution for this second case.

Stem cells and aging

If aging were a degenerative condition as aging itself is origin for degenerative diseases (aging related pathology), then expectations regarding stem cells in treatments of degenerative conditions should advance also towards treatment of aging as such. Under these conditions we had to answer the very difficult question „what is aging?" and how important alterations of stem cells' functions in aging are. The above biochemical hypothesis of aging has responses in the evolutionary context to this question [14]. Anyway, in mammals, aging appears not to result from stem cells' replicative exhaustion which should take place after completion of the maximum lifespan, in the case of humans, at the least. However, there seems to a functional impairment of this type of cells [22]. For instance, in the case of skin epithelial cells there were some modifications, number of stem cells decreased but as well their differentiation capacity, as shown by loss of melanocytes involved in skin and hair pigmentation. From physiological changes associated with aging, we can assume that there are involvements of some proliferation and differentiation deficits of several types of stem cells of skin, muscle, organs and brain.

In young animals, stem cells divide asymmetrically, thus producing other stem cells and cell lines which are differentiated during tissue's homeostasis and regeneration. Some stem cells lose their specificity with age and produce impaired descendants, a fact which determines loss of tissue's integrity and physiological decline, even the number of stem cells is unaffected. Some stem cells lose their renewal capacity by undergoing symmetrical divisions that result in two differentiated cells and a gradual loss of the stem cell stock. Senescence of stem cells can also contribute to loss of functional stem cells. Increase in malignant phenomena with age, especially epithelia, of a high turn-over has been assumed as originating in the stem cells' compartment or stem cells' early descendants [22].

Depending on tissues' capacities to proliferate and regenerate, there are three types of tissues: tissues with proliferative (turn-over) and high regeneration capacities (epidermis, blood, intestine), which have well represented stem cell compartment; tissues with a more reduced turn-over, but high regenerative capacity (muscle, liver) for which there are different strategies of regeneration and repair such as need for stem cells in the case of muscles and proliferation of differentiated hepatocytes in the case of liver; tissues with reduced turn-over and regenerative capacity (brain, heart) for which repair is limited and takes place through recruiting stem cells [23]. Taking into account the modalities in which various tissues are torn over time, we can detect mechanisms determining aging, levels at which capacities of regeneration and renewal of systems are affected meaning the level of stem cells and cells' impaired differentiation.

Which is the determinant of aging? Is it that cells lose their capacity as stem cells, their capacity to differentiate or do they more poorly differentiate and the resulting cells are less functional?

As regards stem cell functioning, most of the studies were conducted on hematopoietic stem cells (HSC), which can be used for bone marrow repopulation in cases of severe irradiations. Bone marrow transplant gained its utility in clinics at the end of sixties, in cases of acute leukemia and hereditary immune deficiencies. Then it was found out that actually hematopoietic stem cells from the bone marrow migrate to the blood (peripheral HSC) and can restore the hematopoietic and immune functions in patients through stimulation with cytokines or in the recovery stage after myelosuppressive therapy [24].

However, apart from hematopoietic stem cells' utility for the clinics, much is unknown regarding these cells and most knowledge is due to studies conducted on mice. Even though the model of immune suppression in mice is not similar to that of bone marrow repopulation through transplantation after lethal doses of radiations, it provides information about the HSC repopulation potential.

From the existing data, the age of the donor appears to be determinant for a successful transplant. In aging it seems that the ratio adipose cells to hematopoietic cells of the bone marrow [24] changes and more so, the number of CD34+ cells (having a specific antigen that has immune function) decreases. A modality to solve the problem of a low number of hematopoietic cells in the case of these grafts is cells' multiplication in cultures and re-implantation. Nonetheless, it appears that the multiplication potential of CD34+ cells is limited and data are controversial. Umbilical stem cells CD34+ can be multiplied only for a limited time, a fact suggesting their "aging"

Generally, HCS could provide interesting information about adult stem cells' dynamics. Also, in the case of HSC (being nevertheless reserved due to controversial results and manipulation techniques), changes in multiplication and differentiation capacities are found out. Which would be the factors involved in these changes?

There are often discussions about the influence of microenvironment on stem cells. Recent data have shown that the environment could determine transdifferentiation without transition involving stem cells. Neurons were produced from skin cells (fibroblasts) subsequently to manipulations [25]. Under these conditions we can hope that through control of the environment, transdifferentiations can be obtained more easily, cheaper and by avoidance of ethical aspects which stem cells raise. In the case of these cells, apart from the success mass media distributed, there were as well many failures familiar to those working in the field. At the same time new perspectives open up in cell therapy, similar to the heart reconstituted from the fibrous skeleton (which resulted from chemical removal of cells) by use of repopulation with cardiomyocytes from new born rats [26]. The rat heart thus formed started to beat eight days after growth under laboratory conditions and stimulation by electrodes. It is interesting how these various tissues formed (blood vessels) through these cell differentiation as if the fibrous skeleton transmitted certain signals. The physical and chemical environment appears to be decisive in the case of any cell, regardless its advanced differentiation. Cells are open systems. Complex influences the environment exerts on cells are nevertheless least known.

Another aspect would be that of the environment of differentiated tissues, not only that of stem cells. Cells from differentiated tissues (or host tissues in the case of grafts with stem cells) could change biochemically with age, and this fact might lead to limiting differentiation of stem cells reaching those tissues or of stem cells responsible for regeneration of aforementioned tissues. For instance, HCS can lead to regeneration of tissues outside of the hematopoietic bone marrow. The phenomenon of differentiation of stem cells transplanted into other types of cells than the desired type was already mentioned. According to predictions of the biochemical hypothesis of aging, the above phenomenon might be treated through transcription factors eventually having local action and promoting stem cell differentiation. With a view to restoring the initial cellular biochemical features of stem cells from different tissues, epigenetic imprinting of tissues from young ages could play an orienting role.

Future researches could have answers in this sense, thus contributing to deciphering the aging process. Despite all difficulties stem cells' research encountered, it has shown promising ways for cell therapy emerging as important part of regenerative medicine.

References

1. Stambler I, (2015). Stop aging and disease, ICAD 2014, conference report, Aging and Diseases,Vol. 6, No. 2, pp. 76-94.
2. Banerjee S.,Williamson D., Habib N., Gordon M., Chataway J., (2011).Human stem cell therapy in ischaemic stroke: a review, Age and Ageing no. 40 pp. 7–13
3. Yamazaki S., Iwama A., Takayanagi S., Morita Y., Eto K., Ema H. Nakauchi, H.. (2006). Cytokine signals modulated via lipid rafts mimic niche signals and induce hibernation in hematopoietic stem cells, EMBO J., vol. 25, pp. 3515–3523.
4. Boyer L.A., Lee T.I., Cole M.F., Johnstone S.E., Levine S.S., Zucker J.P.,et al. (2005). Core transcriptional regulatory circuitry in human embryonic stem cells[1]. Cell, vol. 122 no. 6, pp. 947–56. doi[2]:10.1016/j.cell.2005.08.020[3]. PMC[4] 3006442[5]. PMID[6] 16153702[7].
5. Almeida-Porada G., Porada C.D. (2004). Formation of human hepatocytes by human hematopoietic stem cells in sheep, Blood, no. 104, pp. 2582-2590.
6. Abad M. A., Mosteiro L., Pantoja C., Cañamero M., Rayon T., Ors I. et al. (2013). Reprogramming in vivo produces teratomas and iPS cells with totipotency features. Nature, vol. 502, no. 7471 pp. 340–5. doi[8]:10.1038/nature12586[9]. PMID[10] 24025773[11]

---

1. https://www.ncbi.nlm.nih.gov/pmc/articles/PMC3006442
2. https://en.wikipedia.org/wiki/Digital_object_identifier
3. https://dx.doi.org/10.1016%2Fj.cell.2005.08.020
4. https://en.wikipedia.org/wiki/PubMed_Central
5. https://www.ncbi.nlm.nih.gov/pmc/articles/PMC3006442
6. https://en.wikipedia.org/wiki/PubMed_Identifier
7. https://www.ncbi.nlm.nih.gov/pubmed/16153702
8. https://en.wikipedia.org/wiki/Digital_object_identifier
9. https://dx.doi.org/10.1038%2Fnature12586
10. https://en.wikipedia.org/wiki/PubMed_Identifier

7. Takahashi K. , Tanabe K., Ohnuki M., Narita M., Ichisaka T., Tomoda, K. Yamanaka, S. (2007). Induction of Pluripotent Stem Cells from Adult Human Fibroblasts by Defined Factors. Cell Vol. 131, no.5 pp. 861–872.doi[12]:10.1016/j.cell.2007.11.019[13]. PMID[14] 18035408[15].
8. Yu J., Vodyanik M. A., Smuga-Otto K.,Antosiewicz-Bourget J., Frane J. L., Tian S., Nie J. et al. (2007). Induced Pluripotent Stem Cell Lines Derived from Human Somatic Cells Science , vol. 318, Issue 5858, pp. 1917-1920.
9. Moore,K.L.,Persaud T.V.N.,and Torchia M.G.(2013).Before we are born:Essentials of Embriology and Birdh Defects,Philadelphia,PA:Saunders,Elsevier Print.
10. Hochedlinger K.,Yamada Y., Beard C., Jaenisch R.(2005).Ectopic expression of Oct-4 bloks progenitor-cell differentiation and causes dysplasia in epithelial tissues.Cell,vol121,no.3,pp465-77.
11. Ibrahim E.E.,Babaei-JadidiR.,Saadeddin A.,Spencer-Dene B.,Hossaini S.,Abuzinadah M.,(2012).Embrionic Nanog activity defines colorectal cancer stem cells and modulates through AP1-and TCF-dependent mechanisms. Stem cells ,vol.30,no10,pp2076-87.
12. doi[16]:10.1002/stem.1182[17].PMID[18] 22851508[19].
13. Santagata S.,Hornich J.L.,Ligon K.L.(2006).Comparative

11. https://www.ncbi.nlm.nih.gov/pubmed/24025773
12. https://en.wikipedia.org/wiki/Digital_object_identifier
13. https://dx.doi.org/10.1016%2Fj.cell.2007.11.019
14. https://en.wikipedia.org/wiki/PubMed_Identifier
15. https://www.ncbi.nlm.nih.gov/pubmed/18035408
16. https://en.wikipedia.org/wiki/Digital_object_identifier
17. https://dx.doi.org/10.1002%2Fstem.1182
18. https://en.wikipedia.org/wiki/PubMed_Identifier
19. https://www.ncbi.nlm.nih.gov/pubmed/22851508

analysis of germ cell transcription factors in CNS germinoma reveals diagnostic utility of NANOG.. The American Journal of Surgical Pathology, vol.30,no12,pp.1613-1618.

14. Kregel S., Kiriluk K.J.,Rosen A.M.,Cai Y.,Reyes E.E.,Otto K.B.,et al.(2013).Sox2 is an androgen receptor-repressed gene that promotes castration-resistant prostate cancer.PloS ONE8(1):e53701.doi:10.1371/ journal.pone.00537010.PMC3543364.PMID 23326489.
15. Diaconeasa AG, Rachita M.,Stefan-van Staden R-I.,(2015).A new hypothesis of aging,Med Hypothesis no.84,pp252-257.
16. Beckman J.,Scheitza S.,Wernet P.,Fischer J.C.,Giebel B (2007).Asymetric cell division within the human hematopoietic stem and progenitor cell compartment: identification of asymmetrical segregating proteins. Blood vol.109,no12.,pp 5494-501.doi:10.1182/ blood-2006-11-055921.PMID 17332245.
17. Xie T.,Spradling A.C.(1998).Decapentaplegic is essential for the maintenance and division of germline stem cells in the Drosophila ovary.Cell 94(2):pp.251-60.doi:10.1016/ S0092-8674(00)81424-5.PMID9695953.
18. Song X.,Zhu C.H.,Doan C.,Xie T.(2002) Germline stem cells anchored by adherens junctions in the Drosophila ovary niches.Science,Vol 296 no5574,pp.1855-7.
19. Diaconeasa A.G.,Rachita M.,Dumitrescu I.,(2013) A new hypothesis of the origin of menopause,Romanian J.og gerontology and Geriatrics ,no.1,pp.19-24.
20. Lindvall O.,Kokaia Z.,Martinez-Serranoa.,(2004) Stem cell therapy for human neurodegenerative disorders-how to make it work, Nature publishing Group,http://www.nature.com/ naturemedicine.
21. Mimeault M.,Batra S.K.,(2009) Recent progress on tissue-resident adult stem cell biology and their therapeutic

implications, Stem Cell Rev 4 (1), pp. 27-49.

22. Bakhuraysah, M. M., Siatskas, C., & Petratos, S. (2016). Hematopoietic stem cell transplantation for multiple sclerosis: is it a clinical reality? Stem Cell Research & Therapy, 7, 12. http://doi.org/10.1186/s13287-015-0272-1
23. Liu L., Rando T, (2011), Manifestatios and mechanisms of stem cell aging, J.Cell.Biol,,vol.193 no.2, pp. 257-266.
24. Rando T.A. (2006). Stem cells, ageing and the quest for immortality, Nature, no 441, pp.1080-1086.doi10.1038/nature04958
25. Ho A.D.,Punzel M. (2003), Hematopoietic stem cells; can old cells learn new trics? J.Leukoc. Biol no. 73 pp.547-555.
26. Pang Z.P., Yang N., Vierbuchen T., Ostermeier A., Fuentes D.R., Yang T.Q.et al., (2011). Induction of human neuronal cells by defined transcription factors. Nature, Vol. 476 no.7359, pp. 220-3. doi: 10.1038/nature10202.
27. Otti H., Matthiesen T., Goh S. et al. (2008), Perfusion-decellularized matrix: Using nature's platform to engineer a bioartificial heart. Nature Medicine.;doi:10.1038/nm1684.

A UNIFIED VIEW ON DEGENERATIVE DISORDERS

Introduction

The pharmaceutical industry currently finds itself in a great crisis of ideas that has been around for decades. Progress in medicine is limited to a few subfields, such as infectious diseases, through the discovery of antibiotics, the development of vaccines etc. The cardiovascular revolution has raised life expectancy in the developed West by 20 years, while traumatology has benefited from the discovery of pain killers. But these days society is confronted more and more with degenerative illnesses, as much due to the increase in life expectancy and the average age in the population, as to the lifestyle change, leading to exposure to risk factors such as sedentary life, stress, and overeating.

The current strategy for understanding degenerative illnesses is looking for genes that may be responsible for them, a strategy often criticized by philosophers of science (1). It is completely normal for this to be the case, given that recognizing the role of epigenetic modifications, which precede mutations in illnesses such as cancer, was considered heretical a few decades ago. What is happening in the present is just a continuation of this way of thinking.

Instead of searching for genetic predispositions for diseases like cancer, it is more important, in terms of prevention, to determine which environmental factors trigger the illness. In order to develop efficient treatments, we might add, it would be important to know how the environment influences genes, whose expression may undergo modifications, and also to understand other regulating elements of biochemical cellular reactions. Looking for genes for some degenerative diseases such as type II diabetes or conditions like obesity has not lead to a better understanding of the illnesses or to more efficient treatments. Genes have been identified that are relatively significant in triggering type II diabetes, for example, but none of them plays a crucial role. Furthermore, they differ among different human populations (2).

Cases where the mutation of a gene, which is then transmitted in the classic Mendelian fashion, is responsible for a degenerative disorder, are rare. They are the exception, not the rule. In general, most degenerative illnesses have a hereditary component, but they are polygenic and their transmission is not simple, but subjected to numerous expression regulations, as well dependent on the availability and the activity of the proteins (enzymes) they codify. In other words, up until now, the discovery of mutations at the origins of certain hereditary conditions has been a series of lucky strokes for researchers. Now, we find ourselves traversing a time period in which there probably is need for greater efforts, especially in understanding, in order to get to the bottom of what causes degenerative diseases, and how their mechanisms work. Put in a different way, the luck of finding a monogenic, simple deterministic cause for degenerative illnesses has passed. This has to be the reason for which progress in treating these conditions has been so modest in the last few decades. There is need for new paradigms and approaches. In what follows we will sketch one such unified approach to degenerative illnesses, based on a new hypothesis on aging (3).

Towards a new classification of degenerative illnesses

What are degenerative diseases? What is, in fact, degeneration? What is happening in such a condition? What tissues can undergo degeneration? How many types of degenerative illnesses are there? We will try to answer all of these questions in turn.

Degenerative diseases are those conditions in which a tissue or organ loses its initial function, or said function is diminished. Degeneration appears very early, in the embryonic stage and continues throughout childhood, but, when it does not lead to modifications in adaptation, we consider this type of degeneration a part of the organism's development. When occurring during these stages of development, degeneration can be a source of evolution, leading to modifications in some functions. It is, however, usually considered pathological after the organism has developed up to a limit that is considered normal, or at least up to the point of birth and possibly during the first few years of life. For example, progeria, a group of syndromes characterized by accelerated aging, of which the best known is Hutchinson-Gilford Progeria progeria, appears during the first years of life. There are autoimmune conditions and mental disorders (such as bipolar disorder, schizophrenia, and autism) that also appear during childhood. Aside from those, there are illnesses that severely affect a certain tissue and jeopardize the life and development of children, and are often lethal, but they are treated mostly as genetic illnesses, not degenerative ones. In order for them to be named "degenerative", most of them have to appear in old age, and indeed, old age is when a veritable explosion of degenerative conditions occurs. Biochemical reactions that occur during the first years of life or before birth and affect the child's chances for survival and normal development are taken to be causes of genetic defects. There is an intellectual transfer of the importance given to genes to the case of illnesses that develop during adult and especially old age. But in the latter case there is an important approach to their treatment, which is viewing them through the lens of metabolic modifications and biochemical alterations. Perhaps a reverse transfer is needed, according to which disorders appearing during the first years of life and before birth would be viewed as accelerated biochemical alterations. This phenomenon could help create a change in the therapeutic approach.

Take, for example, Hutchinson-Gilford progeria. It occurs in childhood and it is considered to be accelerated aging because of the patients' phenotype, which is similar to that of old people. Even this case is in need of a more nuanced perspective. One of the genetic defects that are present in parents of children with progeroid syndromes (4), but, even though the cellular alterations that are present in the children have been detected in their parents (to a lesser degree), the parents are healthy.

The Down syndrome, whose genetic determination is very clear, is considered to be an accelerated aging syndrome, with premature hypercholesterolemia, metabolic syndrome, and atherosclerosis. Where other diseases are concerned, such as dominantly transmitted diseases with incomplete penetrance (not all carriers get the disease, but about 50% of them do), like some forms of schizophrenia, there are homozygote twin pairs in which one is sick and the other healthy. The epigenetic component is the most important in the case of illnesses of this type.

In general, the prevailing view on degenerative pathology (which we would dub endogenous biochemical pathology) is that any debilitating disease that affects the chances for survival and development before birth and during childhood is a fabrication defect (meaning, genetic). Later on, when the organism is mature, it is considered degradation (degeneration). Thus, this is a mechanistic view, wherein a warranty period is postulated, during which the organism (the machine) should function autonomously, without any important reparations, and any major dysfunction is due to a fabrication defect. Eventually, after the warranty period is over, it is considered that the organism degrades due to use.

In reality, organisms are not machines, but matrices of biochemical reactions. In chemistry, even though entropy is present, there is no friction or mechanical forces that degrade the materials. Although living organisms have mechanical parts subject to such processes, in chemistry reactions are reversible. Of course, in biochemistry, though many reactions are reversible, there are limitations given by the availability of certain enzymes. Biochemistry makes living organisms capable of regenerating some structures, at least partially. A biochemical approach to degeneration that is as unified as possible would be very welcome. This means that illnesses with degenerative characteristics that appear early in life can borrow from the metabolic approach to aging pathology, while aging, with all the illnesses it is associated with, can borrow from the view on pathologic mechanisms. Under the light of this approach to degenerative maladies we should also include some "classic" genetic illnesses, with a clear genetic determining mechanism, and aging, with all its associated pathologies, including effects like wrinkles, hair loss and pigmentation loss, reproductive decline, and especially menopause in women. This last typically human character and its appearance in evolution will be discussed at length in another article.

In other words, aging itself should be treated as a special disease, with multiple complications, which represent its pathology. Unfortunately, the reticence regarding viewing aging as an illness persists. Only recently there has been research published that tries to impose this idea (5), even though Ilia Metchinkov, the founder of gerontology, and later Ana Aslan, who created the first institute of geriatrics in the world, have opened the way to such an approach.

From the point of view of potential treatments, it may be considered lucky that few such diseases have a simple and clear genetic determinism. In general, these "classic" genetic illnesses are the point one reaches when there are no alternative, metabolic ways of avoiding the specific metabolic deficiency. Given the degree of redundancy we can observe in biology, most of the time there exist several ways of obtaining the same cellular reaction product or physiological function. In addition, theoretically, epigenetic modifications can be controlled by blocking or activating certain genes.

This approach could lead to simpler ways to find treatment for rare diseases, or for those with high incidence in populations from poor areas of the world, which makes investing in their research not profitable.

In these circumstances, we can ask ourselves: what does it mean to be perfectly healthy? What is the reference frame in this case? Traversing individual biological limits, and species limits, could represent the new stage in the evolution of the species in question.

What is it that goes on in degenerative maladies? Oftentimes structures become diminished or even disappear, as happens in the case of the regression of some glands. This is sometimes considered physiologically normal, not only during development, as is the regression of the thymus gland, but also later on. For example, the number of melanocytes in the skin decreases with age. The cerebral ventricles are enlarged in Alzheimer's disease (6), which might be connected to the loss of some neurons from the periventricular areas (sources of neuronal stem cells), which would lead to the slowing down of neuron replacement, but also from other areas. Ventricle enlargement has also been observed in schizophrenia (7) and other brain disease like autism (8). Other times, one type of cell is replaced by another, as in hepatic or renal fibrosis.

There are several types of organ degeneration (e.g. fatty, fibrous). Sometimes, accumulations of extracellular substances occur, like atheromatous and amyloid plaques. Cellular degeneration is given by apoptosis, cellular senescence (the irreversible cessation of cell division) that appears *in vitro*, but whose existence *in vivo* is controversial, dysplasia, in which cell maturation and differentiation are delayed (e.g. epithelial dysplasia), metaplasia[1], in which cells of one mature, differentiated type are replaced by cells of another mature, differentiated type. In addition, oftentimes some cells and cell components are attacked by the immune system.

1. https://en.wikipedia.org/wiki/Metaplasia

The tissues that suffer degeneration are the skin, the arteries, the glands (the thyroid, the pancreas, the gonads, the adrenals, the hypophysis), the brain, the immune system, the intestines, the kidneys, the liver, the heart, and the bones and muscles as well. But most often, the epithelial tissues and endothelia suffer degeneration, being cells that divide rapidly, and have a role of protection, nourishment, and which have an intense metabolic activity (the glands). A large number of degenerative illnesses are autoimmune diseases (multiple sclerosis, amyotrophic lateral sclerosis, type I diabetes, lupus erythematosus, rheumatoid arthritis, psoriasis, vitiligo, alopecia areata, etc). Aside from the numerous autoimmune illnesses, autoimmune and/or inflammatory processes may accompany not only aging in general, but also illnesses associated with aging, such as Alzheimer's, and diseases like autism, schizophrenia, depression, in which an allergy to inflammation factors may be present (9). Moreover, in animal models of Alzheimer's, suppressing the immune system has lead to the restoration of memory (10). Likewise, autoimmune or inflammatory phenomena may be involved in conditions like menopause (degeneration of the ovaries) and the degeneration of the thyroid. Women with autoimmune diseases reach menopause younger (11). A form of autoimmune alopecia was cured by a drug that blocks the immune system (12). In fact, it is a drug used in cancer, and some of these drugs may have cellular regenerative effects (3). We consider that this drug or others with similar mechanisms could be used in other similar conditions.

A new, unified approach to degenerative diseases

In what follows we will present a unified approach to degenerative diseases, showing that the phenomena involved in degeneration have a common mechanism, explained by the biochemical hypothesis of aging, which applies in this case as well (3).

First of all, cellular and biochemical modifications actually share the same type of phenomenon: the diminution of some cellular reactions and their replacement with other reactions, in agreement with the biochemical hypothesis on aging (3). This phenomenon is reflected in the differentiation of stem cells as well, which may lead to different types of cells, such as skin cells, which lead to more types of cells, among which melanocytes, likewise immune system cells, other epithelial glandular cells, but very likely neuronal cells as well. The stem cells simply no longer have the necessary instructions, because of modifications in their biochemical environment, to differentiate into all types of cells. This is a well known phenomenon. Stem cells harvested from old individuals have a smaller capacity for differentiation, while, with age, some types of cells become much more homogeneous. An argument for this idea are the experiments with *in vitro* fibroblasts, which show that after being harvested and deprived of nutrients (serum), "old" cells in the stationary phase secrete more collagen, not in absolute numbers, but in relative numbers, meaning to the detriment of other proteins (13). They do not synthesize more collagen, but they stop synthesizing other proteins. The fact that aged fibroblasts synthesize collagen instead of other proteins (though wrinkles and stretch marks are collagen ruptures) suggests the forming mechanism of fibromas, keloids, and pigment moles in black people (with African descent). Collagen is the target of autoimmune attacks in many cases, among which lupus erythematosus and rheumatoid arthritis.

We consider that this experimental model (cultivated cells in stationary phase, after their proliferation has ceased), is a good model for what happens in aging *in vivo*. It is well known that cultivated cells suffer a loss of metabolic specificity, and experiments we conducted show that their response to stress is modified (3). Alterations in the cytoskeleton seem to be responsible for this phenomenon that manifests itself through modifications in cellular contractions. Indeed, this phenomenon seems to have an *in vivo* correspondent. Modifications occurring in the cardiomyocyte cytoskeleton, through the accumulation of vinculin in certain cellular zones, seem to play an adaptive role in the case of heart aging (14). When certain reactions lack, those that remain are overwrought, when there is an excess of energy, which determines chain reactions with remote alterations of other reactions (3).

It is well known that a high density of cells represents a risk of malign tumors, not only because they become difficult to detect, but also because the odds of a tumor forming are increased. A high density of cells could also be, in the case of the growth of an organism, the cause of the "dilution" of intercellular signals, which could then lead to a weaker stimulation of some reactions, visible through the number of cell receptors, a similar phenomenon to what happens in insulin resistance. For this reason, cell cultures can furnish many more clues about aging than we might think at first sight. This might also explain the effects of some anti-aging treatments, of which it is known that they work, but which do not have a satisfying scientific explanation (e.g. peeling). The effects of meiosis could, at least partially, be explained in the same way.

Aging is, in the first place, the degeneration of supporting tissues and with a role in nourishment (3). At least these are the phenomena we notice first. As has already been observed, aging occurs at different rates in different tissues (3). If aging or the degeneration of tissues occurs because some reactions do not take place anymore or are diminished, then the diminution through antagonists of some hormones or mediators, substances with a role in extracellular signaling, and the systematic blocking of some receptors with a role in intracellular signaling, should lead to a faster aging process in some tissues. Indeed, this is the case. antidepressants or anti-allergy drugs with anticholinergic action significantly increase the risk of Alzheimer's, in which the cholinergic system is affected (15) Progesterone antagonists, including cortisol, have negative effects on the ovaries, influencing the age of the onset of menopause, and so do some pollutants that act in the same way as progesterone antagonists (16)

The opossum ages very quickly in conditions of stress (17). Stress accelerates aging in other species as well, including mammals, which could be explained, at least partially, through the blocking of numerous specific cellular reactions under the influence of cortisol and other hormones. But stress, like any other stimulus, has effects that vary according to dose and way of administering, and the long-term effects can be surprising. Although telomerase, probably one of the syntheses blocked by cortisol, decreases in conditions of stress, in the long term it increases (18), which might explain some cases of longevity in people subjected to extremely harsh conditions, like extermination camps. The hormesis (moderate stress) effects are in accord with this approach (3). In fact, stress can activate long-term reactions which slow down aging.

In light of the same hypothesis, a higher level of some specific cellular reactions (characteristic of some types of cells, the mark of differentiation) should, through exogenous stimulation or simply because of some genetic factors, have protective effects against aging/degeneration. In this sense, we can note the effects of physical activity on delaying aging and other degenerative illnesses. During physical exercise, although stress hormones are released, a lot of reactions are being stimulated, especially those involved in energy metabolism (mitochondria are even created *de novo*) (19), but also others, including immune reactions (20). Physical activity decreases insulin resistance.

Likewise useful in preventing the effects of aging and degeneration are intellectual preoccupations. The risk of developing Alzheimer's decreases in direct proportion to a person's education. Intellectual pursuits have the effect of delaying the onset of the illness, but in intellectuals who get it, the illness has a sudden development (21), which is suggestive of the cellular dynamic of the apparition of the disease

Stimulating the brain, through its connections to the endocrine system and not only, could have stimulating effects on many specific cellular reactions, including the repair reactions. This may be the answer to why Stephen Hawking, whose life expectancy after the onset of his illness (amyotrophic lateral sclerosis) was a few years, survived for decades. Another link has recently been discovered between the brain, the central nervous system, and the immune system, using the lymphatic system as a medium (22).

The high level of some cellular reactions could influence the cellular metabolism as a whole, especially energetically, with effects on other reactions, especially the ones involved in repair, but also other syntheses. It is known that the skin of African people generally ages slower than that of Caucasians. What is more, in the case of people of African decent, the percentage of people who look a lot younger than they are (by 10 years or more) is twice as high as that of Caucasians (10% vs. 20%). Photoaging is considered to be the main cause for rapid skin aging in Caucasians. But the African skin has other characteristics as well, such as a superior elasticity, due to more fibroblasts and a specific collagen which is less prone to photoaging. Some in vitro studies report remarkable differences regarding stratification and differentiation according to skin types, as demonstrated by histological appearance, gene expression analysed by DNA microarray and quantitative proteomic analysis (23). The authors noticed that African epidermis displayed more dividing cells but less terminal differentiation, which suggests, they pointed our, a different epidermal programs in each skin type, in particular at the level of differentiation. With other words, there is another pattern of cell proliferation and differentiation.

In this conditions, can we activate this program of skin cell differentiation in other ethnic groups? Although we know than even the pigmentation can be improved in some conditions. Likewise, there are Caucasians who have suffered severe sunburns and continue to look much younger. The genotype of these individuals might be what offers them a greater capacity for DNA repair, but research results to appear in the "International Journal of Dermatology" show that skin cells of individuals who look much younger have a greater capacity for synthesizing various substances, wax, nutrients, lipids, with a role in hydration (24). Wrinkles are not a necessary sign of aging, because some people, including some very pale Caucasians, do not develop them during aging. A better response to stress, better reactions, protects against what we call aging. Can be apply this idea to other traits?

The highest number of answers about degeneration could come from human evolution itself. The human is an animal that suffers from more degenerative illnesses than his closest related species (2). H. sapiens suffers from autism, schizophrenia, other mental illnesses, Alzheimer's (senile dementia), autoimmune disorders etc. Furthermore, the incidence of type II diabetes, cancer, and other degenerative diseases is the highest in humans. Aging in humans is accompanied by a specific pathology and is more degenerative and disabling. The incidence of these diseases, which are rare to non-existent in chimpanzees (in captivity), has greatly increased after industrialization. The human aging pathology also includes an abrupt reproductive decline, the female menopause. Also, some diseases which are very severe in humans, are a lot less serious in chimpanzees, if they do get ill. Chimpanzees do not have AIDS. All attempts to infect chimpanzees with HIV have failed. Although it does not seem to be due to an increased resistance which is connected to specific antigens, we can take this aspect into consideration. But the Simian Immunodeficiency Virus (SIV), though it produces the illness, its severity is incomparable to that of AIDS. Regarding the hepatitis B and C viruses, even if they do produce infections, the hepatitis they cause is not accompanied in the chimpanzee by the same serious degenerative effects they give in humans.

These phenomena could be attributed to different forms of siglecs (glycoproteins on the cell surface that regulate immune system activity). It may be said that these act like brakes on the immune response, somewhat weaker in humans (2). But the question becomes, why did these defective siglecs appear, what is their role? On the basis of many metabolic, genetic, neurological, and paleontological analyses, we have advanced the hypothesis that human evolution could be the result of an alternation between periods of scarcity and relative abundance of resources. Adapting to few resources may have caused epigenetic, hereditarily transmitted modifications, which persisted even after the resource crisis was over. The IGF1 (insulin-like growth factor 1)/insulin pathway, one of the main targets of caloric restriction, would have suffered activation and regulation modifications. Indeed, this pathway, especially the enzyme with Akt regulatory activity (protein kinase B), is affected in diabetes, cancer, Alzheimer's, mental illnesses, autism. Any typically human mental illnesses that appears in childhood (autism), youth (schizophrenia or affective psychoses) or old age (Alzheimer's) can be considered a kind of diabetes or hypoglycemia of the brain (2). This topic will be discussed further in another book.

To the question of whether this pathway is affected in autoimmune disorders as well, the answer is given by the fact that rapamycin, a blocker of the immune system used in transplants, which prolongs life in rats (3), acts exactly on this pathway (with inhibitory effects).

Humans have approximately the same metabolic rate as the chimpanzee (*Pan troglodytes*), but a brain which is 3 times bigger and takes up 20% of the resting metabolic rate. The insulin levels are similar for the two species (25), but the level of IGF1, which acts similarly to insulin, but has mitogenic effects, is about 6 times less in the human than in the chimpanzee. There is a connection between IGF1 and body mass, but the mass difference between humans and chimpanzees does not justify this difference. Although some organs are smaller in humans than their cousins (the intestine, muscles), in general, excluding the brain and the placenta, human cells have a lot less energy at their disposal than do chimpanzee cells, which means fewer specific syntheses. This effect is easy to observe, including in the difference in strength between the two species. This difference may have been at the origin of some typically human characteristics (2).

Precisely because individual cells have at their disposal a smaller quantity of energy to synthesize specific proteins, in cases in which the homeostasis of some substances dictates it (a certain level of hormones, as in the case of diabetes or the Polycystic Ovary Syndrome (PCOS), numerous pancreatic or ovarian cells are forced to proliferate, differentiate, or secrete certain products in large quantities. Resistance to insulin leads to insulin synthesis in large quantities by the pancreas, which exhausts the gland, resulting in incapacity to synthesize insulin. Something similar, we think, happens in the case of PCOS ovaries.

But this phenomenon seems to appear in other conditions, too. Contrary to that it may look like at first sight, in schizophrenia there is documented a fronto-cortical dopamine hypoactivity and a striatal hyperactivity (26), although the picture seems much more complex. Still, antipsychotics act and have a positive effect by blocking dopamine receptors (dopaminergic activity). Probably, in this disease, but also in others (epilepsy and not only), precisely because individual neurons have a smaller capacity for synthesis and response to certain signals, supported by new data (26), it comes to a concert response from many cells. A low energy and a less activity of the IGF1 pathway is documented in schizophrenia (2).

This phenomenon, the decreased cell energy, can lead to premature proliferation and defective differentiation of specialized stem cells, which has an effect on the stock of stem cells, regarding their number and their future capacity for differentiation as well. A similar process may be taking place in PCOS, which is often associated with insulin resistance (27) and with premature menopause. But it would probably take place in many tissues and organs, especially those where cells are forced to divide rapidly.

Cells which divide rapidly, without being sufficiently "ready", can propagate errors, which first of all leads to apoptosis, and then, very probably as a way of adapting, to cellular transformation (malignization). Women with mutations of the BRCA1 and BRCA2 genes, involved in checking the errors in cell division, aside from having an increased mammalian and ovarian cancer, reach menopause earlier (28).

Even in Alzheimer's, there are data from animal models that suggest that neuronal cells, when they try to divide, suffer apoptosis (29). This phenomenon is in agreement with viewing Alzheimer's as a diabetes of the brain (2). In this illness, neuronal stem cell proliferation and differentiation would be affected (2). When cells dispose of little energy there are high chances that they will suffer apoptosis. Cells that divide rapidly while not energetically and signaling equipment-wise (cellular receptors, capacity for synthesis) prepared, they can suffer apoptosis, which triggers reactions of the immune system. The remains of cells that have suffered apoptosis are essential for attracting the immune system cells which trigger the autoimmune response. A murine animal model shows that, if the remains are removed, the autoimmune response does not occur (30).

Metformin, a drug used in diabetes to reduce insulin resistance, has therapeutic effects in PCOS (27). At least in some cases, insulin resistance may be caused by an inherently lower capacity for the synthesis of certain metabolites, for responding to extracellular signals, as in the case of PCOS, when a diminished capacity to synthesize progesterone is manifested, which cannot be restored with metformin (27). Diabetes, a model of rapid aging, could furnish more information about aging, aside from the effects tied to the insulin pathway and the complications of this malady.

Insulin resistance could sometimes start from an incapacity not only to respond to insulin (to encapsulate glycogen in muscles and the liver), but also to synthesize other metabolites, which would reduce the need to introduce glucose into tissues. Metformin blocks glucose synthesis in the liver, which leads to a decrease in insulin resistance.

The reduced capacity to perform specific syntheses can be, in some cases, the source of insulin resistance, which then, through the reduced contribution in cellular energy that can affect other syntheses, would lead to an increase in insulin resistance. Of course, this phenomenon would also affect the immune system, whose response would become more non-specific (autoimmunity) and disproportionate.

Conclusions

First of all, in order to treat degenerative illnesses and aging, we need new markers of these processes, which reflect the real and relevant modifications occurring with age. Most markers of aging and degenerative diseases, few as they are, also have moderate predictive value. An important example would be the amyloid plaques present in Alzheimer's, which do not proportionally reflect the cognitive deficit (31). The markers we have proposed on other occasions as well, such as the protein composition of cells, the capacity to accumulate and degrade macroergic substances such as glycogen or ATP. On the other hand, some reverse engineering would be needed to see more precisely what are the first reactions and cellular functions affected by aging and degeneration. Judging by the way lipofuscin accumulations and atheromatous and amyloid plaques appear, it would seem that the first one to be affected (even if it does not automatically lead to the loss of functions) is the degradation of proteins, lipids, cellular sugars, meaning the catabolic function. This would give the difference between starvation, which would lead to insulin resistance, and caloric restrictions (all the necessary nutrients, but reduced energy), which would reduce insulin resistance, at least in some species. Caloric restriction would stimulate the degradation processes, the processes of producing energy through catabolism. When the instruments to measure them no longer exist, the intake of energy leads to uncontrolled proliferation and accumulations of metabolites, as we see to be the case in degenerative illnesses, and in aging more generally.

This would be the point to discuss the difference between degeneration as loss of cells and degeneration as destruction of the cell function, which affects many cells that survive. Through studies on long-lived people, we have observed that in order to reach a high number of years of life, of course within the confines of a necessary intake of nutrients, it is important to reduce the loss of cell function (proliferation, metabolite accumulations). Precisely because of redundancy, it is possible to survive with fewer cells than we have. For instance, it is possible to survive with 10% of the kidneys. Only after the loss of a maximum allowed quantity of cells is the function of an organ compromised, if the existent cells are functional. But loss of function affects existent cells, even if their number is not greatly reduced. This phenomenon is visible now in societies in which famine does not jeopardize survival anymore. The degeneration observed now may be the secondary effect of crisis situations in which the long-lived people of today would not have survived.

The predisposition to certain degenerative illnesses that put life in danger explains the difference in life duration between the sexes, which is maintained in most known species. Beyond the level of some antioxidant enzymes, the male sex has a higher level of IGF1, humans included (32). Even if the cellular energy level should be higher in males, the reactions would be faster, and some would be diminished or would disappear faster. But even this case would constitute a phenomenon of responding to the environment throughout the course of evolution, that we could call adaptive, and that has its advantages in certain conditions.

Though the skepticism regarding the treatment of aging and degenerative disease persists, there are initiatives, mainly at the level of NGOs studying aging, to start clinical studies with clinical substances that have real anti-aging effects. One such substance is Metformin. Likewise, rapamycin has proven its effectiveness on rodents (3). The problem is that rapamycin increases the risk of developing type II diabetes. Metformin has positive effects in diabetes, but also in conditions associated with insulin resistance, such as PCOS (27), but it did not restore the ovarian cells' capacity to produce progesterone (27). Also, experiments involving connecting the blood circuits of old and new animals has had the effect of reducing the level of degeneration in old animals (3).

Still, the effects have been modest in certain cases. The explanation would be, according to the biochemical hypothesis of aging, that cells from old animals had already suffered modifications because of the reduced stimulations of certain reactions. They needed a larger quantity of differentiation hormones, which is found in the blood from young animals, but perhaps more specific stimulations were also needed.

The science of rejuvenation, in order to have results, will be a special science, one that takes into account many intercellular interactions, but also other levels of organization in living matter. But the first step would be for living processes to no longer be viewed as static photographs, described by genes, or, according to newer takes, by their activation state. Life means continuous movement, with remote and long-term interactions.

References

1. Hubbard R, Genes as Causes, in Vandana Shiva[2]; Ingunn

Moser[3] Biopolitics : a feminist and ecological reader on biotechnology, London ; Atlantic Highlands, N.J. : Zed Books ; Penang, Malaysia : Third World Network, 1995

2. Diaconeasa A G, Civilizaţia foametei/o altă abordare a umanizării, 2007, Sitech, Craiova.
3. Diaconeasa A G, Rachita M, Stefan-van Staden R-I A new hypothesis of aging, *Medical Hypotheses*, 2015; 84: 252–257
4. Verstraeten VL[4], Broers JL[5], van Steensel MA[6], Zinn-Justin S[7], Ramaekers FC[8], Steijlen PM[9], Kamps M[10], Kuijpers HJ[11], Merckx D[12], Smeets HJ[13], Hennekam RC[14], Marcelis CL[15],

---

2. http://www.worldcat.org/search?q=au%3AShiva%2C+Vandana.&qt=hot_author

3. http://www.worldcat.org/search?q=au%3AMoser%2C+Ingunn%2C&qt=hot_author

4. http://www.ncbi.nlm.nih.gov/pubmed/?term=Verstraeten%20VL%5BAuthor%5D&cauthor=true&cauthor_uid=16825282

5. http://www.ncbi.nlm.nih.gov/pubmed/?term=Broers%20JL%5BAuthor%5D&cauthor=true&cauthor_uid=16825282

6. http://www.ncbi.nlm.nih.gov/pubmed/?term=van%20Steensel%20MA%5BAuthor%5D&cauthor=true&cauthor_uid=16825282

7. http://www.ncbi.nlm.nih.gov/pubmed/?term=Zinn-Justin%20S%5BAuthor%5D&cauthor=true&cauthor_uid=16825282

8. http://www.ncbi.nlm.nih.gov/pubmed/?term=Ramaekers%20FC%5BAuthor%5D&cauthor=true&cauthor_uid=16825282

9. http://www.ncbi.nlm.nih.gov/pubmed/?term=Steijlen%20PM%5BAuthor%5D&cauthor=true&cauthor_uid=16825282

10. http://www.ncbi.nlm.nih.gov/pubmed/?term=Kamps%20M%5BAuthor%5D&cauthor=true&cauthor_uid=16825282

11. http://www.ncbi.nlm.nih.gov/pubmed/?term=Kuijpers%20HJ%5BAuthor%5D&cauthor=true&cauthor_uid=16825282

12. http://www.ncbi.nlm.nih.gov/pubmed/?term=Merckx%20D%5BAuthor%5D&cauthor=true&cauthor_uid=16825282

13. http://www.ncbi.nlm.nih.gov/

van den Wijngaard A[16]. Compound heterozygosity for mutations in LMNA causes a progeria syndrome without prelamin A accumulation. *Hum Mol Genet.*[17] 2006 Aug 15;15(16):2509-22. Epub 2006 Jul 6.

5. Bulterijs S, Hull RS, Björk VCE and Roy AG (2015) It is time to classify biological aging as a disease. *Front. Genet 2015.* **6**:205. doi: 10.3389/fgene.2015.00205
6. Raz N; et al. Regional Brain Changes in Aging Healthy Adults: General Trends, Individual Differences and Modifiers. *Cereb. Cortex* **15** , 2005, (11): 1676–1689.
7. Gaser C[18], Nenadic I[19], Buchsbaum BR[20], Hazlett EA[21], Buchsbaum MS[22]. Ventricular enlargement in schizophrenia related to volume reduction of the thalamus, striatum, and

---

pubmed/?term=Smeets%20HJ%5BAuthor%5D&cauthor=true&cauthor_uid=16825282

14. http://www.ncbi.nlm.nih.gov/
pubmed/?term=Hennekam%20RC%5BAuthor%5D&cauthor=true&cauthor_uid=16825282

15. http://www.ncbi.nlm.nih.gov/
pubmed/?term=Marcelis%20CL%5BAuthor%5D&cauthor=true&cauthor_uid=16825282

16. http://www.ncbi.nlm.nih.gov/
pubmed/?term=van%20den%20Wijngaard%20A%5BAuthor%5D&cauthor=true&cauthor_ui d=16825282

17. http://www.ncbi.nlm.nih.gov/pubmed/16825282

18. http://www.ncbi.nlm.nih.gov/
pubmed/?term=Gaser%20C%5BAuthor%5D&cauthor=true&cauthor_uid=14702264

19. http://www.ncbi.nlm.nih.gov/
pubmed/?term=Nenadic%20I%5BAuthor%5D&cauthor=true&cauthor_uid=14702264

20. http://www.ncbi.nlm.nih.gov/
pubmed/?term=Buchsbaum%20BR%5BAuthor%5D&cauthor=true&cauthor_uid=14702264

21. http://www.ncbi.nlm.nih.gov/
pubmed/?term=Hazlett%20EA%5BAuthor%5D&cauthor=true&cauthor_uid=14702264

22. http://www.ncbi.nlm.nih.gov/
pubmed/?term=Buchsbaum%20MS%5BAuthor%5D&cauthor=true&cauthor_uid=14702264

superior temporal cortex. doi[23]:10.1093/cercor/bhi044[24]. PMID[25] 15703252[26] *Am J Psychiatry.*[27] 2004 Jan;161(1):154-6.

8. Movsas TZ, et al. Autism spectrum disorder is associated with ventricular enlargement in a low birth weight population. *The Journal of Pediatrics* 163(1):73-78, 2013.
9. Brietzke E, Stertz L, Simões Fernandes B, Comparison of cytokine levels in depressed, manic and euthymic patients with bipolar disorder, *Affective disorders*, 2009; 116: 214–217
10. Kaufman A C,. Salazar S V,. Haas L T et al, Fyn inhibition rescues established memory and synapse loss in Alzheimer mice. *Annals of Neurology*, 2015; DOI:10.1002/ana.24394[28]
11. Jones, E.K., Jurgenson, J.R., Katzenellenbogen, J.M. and Thompson, S.C., 2012, Menopause and the influence of culture: another gap for Indigenous Australian women? *BMC Women's Health* 12: 43 http://www.biomedcentral.com/1472-6874/12/43
12. Xing I, Dai Z, Jabbari A e*t al.* Alopecia areata is driven by cytotoxic T lymphocytes and is reversed by JAK inhibition. *Nat Med* 20,1043–1049 (2014). https://doi.org/10.1038/nm.3645
13. Manner G, Cell division and collagen synthesis in cultured fibroblasts *Experimental Cell Research*[29] 1971;65: 49–60
14. Kaushik G, Alice Spenlehauer A, Ayla O. SessionsA O, et al. Vinculin network–mediated cytoskeletal remodeling

23. https://en.wikipedia.org/wiki/Digital_object_identifier

24. https://dx.doi.org/10.1093%2Fcercor%2Fbhi044

25. https://en.wikipedia.org/wiki/PubMed_Identifier

26. https://www.ncbi.nlm.nih.gov/pubmed/15703252

27. http://www.ncbi.nlm.nih.gov/pubmed/14702264

28. http://dx.doi.org/10.1002/ana.24394

29. http://www.sciencedirect.com/science/journal/00144827

regulates contractile function in the aging heart, *Science Translational Medicine* 2015: 7:292ra99

15. Coupland CAC, Hill T, Dening T, Morriss R, Moore M, Hippisley-Cox J. Anticholinergic Drug Exposure and the Risk of Dementia: A Nested Case-Control Study. *JAMA Intern Med.* 2019;179(8):1084–1093. doi:10.1001/jamainternmed.2019.0677
16. Vabre P., Gatimel N, Moreau J *et al.* Environmental pollutants, a possible etiology for premature ovarian insufficiency: a narrative review of animal and human data. *Environ Health* 16, 37 (2017). https://doi.org/10.1186/s12940-017-0242-4
17. Baker ML, Gemmell E, Gemmell RT. Physiological changes in brushtail possums, Trichosurus vulpecula, transferred from the wild to captivity. *J Exp Zool.* 1998 Feb 15;280(3):203-12. doi: 10.1002/(sici)1097-010x(19980215)280:3<203::aid-jez1>3.0.co;2-r. PMID: 9472479.
18. Beery AK, Lin J, Biddle JS, Francis DD, Blackburn EH, Epel ES. Chronic stress elevates telomerase activity in rats. *Biol Lett.* 2012 Dec 23;8(6):1063-6. doi: 10.1098/rsbl.2012.0747. Epub 2012 Oct 10. PMID: 23054915; PMCID: PMC3497144.
19. Huertas JR, Casuso RA, Agustín PH, Cogliati S. Stay Fit, Stay Young: Mitochondria in Movement: The Role of Exercise in the New Mitochondrial Paradigm. *Oxid Med Cell Longev.* 2019 Jun 19;2019:7058350. doi: 10.1155/2019/7058350. Erratum in: Oxid Med Cell Longev. 2021 Jan 18;2021:9274841. PMID: 31320983; PMCID: PMC6607712.
20. Shepard RJ, Shek PN. Impact of physical activity and sport on the immune system. *Rev Environ Health.* 1996 Jul-Sep;11(3):133-47. doi: 10.1515/reveh.1996.11.3.133.

PMID: 9000304.

21. Hausner L, Frölich L, Gardette V, Reynish E, Ousse P-J, Andrieu S, Vellas B, Scheltens P , Boada M,. Jones RW, Dartigues JF, Frisoni G , Spiru ., Hasselbalch S , Agüera-Morales E , Rodriguez G, Salva A , Michel JP, Stiens G, Salmon E , Pasquier F,. Ribera-Casado JM, Burns A, Frölich L, Touchon J, Robert Ph, Olde Rikkert M, Martinez-Lage P ,Rigaud AS , Cherubini A, Winblad/M. Eriksdotter Jönhagen ., Bullock R, Costa-Tsolaki M, Camus V, Malick A; Regional Variation on the Presentation of Alzheimer's Disease Patients in Memory Clinics within Europe: Data from the ICTUS Study; *Journal of Alzheimer's Disease*, Volume 21, Number 1 / 2010; ISSN 1387-287

22. Schwartz M, Kipnis J. A conceptual revolution in the relationships between the brain and immunity. *Brain Behav Immun.* 2011 Jul;25(5):817-9. doi: 10.1016/ j.bbi.2010.12.015. Epub 2010 Dec 25. PMID: 21187141; PMCID: PMC3074045.

23. Girardeau-Hubert, S., Deneuville, C., Pageon, H. *et al.* Reconstructed Skin Models Revealed Unexpected Differences in Epidermal African and Caucasian Skin. *Sci Rep* 9, 7456 (2019). https://doi.org/10.1038/ s41598-019-43128-3

24. Alexa B. Kimball, Maria B. Alora-Palli, Makio Tamura, Lisa A. Mullins, Chieko Soh, Robert L. Binder, Neil A. Houston, Emily D. Conley, Joyce Y. Tung, Nikki E. Annunziata, Charles C. Bascom, Robert J. Isfort, Bradley B. Jarrold, Raghu Kainkaryam, Heather L. Rocchetta, Dionne D. Swift, Jay P. Tiesman, Kazumi Toyama, Jun Xu, Xianghong Yan, Rosemarie Osborne. Age-induced and photoinduced changes in gene expression profiles in facial skin of Caucasian females across 6 decades of age. *Journal of the American Academy of*

Dermatology, 2017; DOI: 10.1016/j.jaad.2017.09.012[30]

25. Mann GV[31], Crofford OB[32]. Insulin levels in primates by immunoassay. *Science.*[33] 1970 ;169(3952):1312-3
26. Martel JC, Gatti McArthur S. Dopamine receptor subtypes, physiology and pharmacology: new ligands and concepts in schizophrenia. *Frontiers in pharmacology*. 2020 Jul 14*;11:1003.*
27. Maruthini D[34], Harris SE[35], Barth JH[36], Balen AH[37], Campbell BK[38], Picton HM[39]. The effect of metformin treatment in vivo on acute and long-term energy metabolism and progesterone production in vitro by granulosa cells from women with polycystic ovary syndrome. 2014;29:2302-16. doi: 10.1093/humrep/deu187.
28. Lin WT, Beattie M, Chen LM, Oktay K, Crawford SL, Gold

30. http://dx.doi.org/10.1016/j.jaad.2017.09.012

31. http://www.ncbi.nlm.nih.gov/pubmed/?term=Mann%20GV%5BAuthor%5D&cauthor=true&cauthor_uid=4989299

32. http://www.ncbi.nlm.nih.gov/pubmed/?term=Crofford%20OB%5BAuthor%5D&cauthor=true&cauthor_uid=4989299

33. *http://www.ncbi.nlm.nih.gov/pubmed/4989299*

34. http://www.ncbi.nlm.nih.gov/pubmed/?term=Maruthini%20D%5BAuthor%5D&cauthor=true&cauthor_uid=25139174

35. http://www.ncbi.nlm.nih.gov/pubmed/?term=Harris%20SE%5BAuthor%5D&cauthor=true&cauthor_uid=25139174

36. http://www.ncbi.nlm.nih.gov/pubmed/?term=Barth%20JH%5BAuthor%5D&cauthor=true&cauthor_uid=25139174

37. http://www.ncbi.nlm.nih.gov/pubmed/?term=Balen%20AH%5BAuthor%5D&cauthor=true&cauthor_uid=25139174

38. http://www.ncbi.nlm.nih.gov/pubmed/?term=Campbell%20BK%5BAuthor%5D&cauthor=true&cauthor_uid=25139174

39. http://www.ncbi.nlm.nih.gov/pubmed/?term=Picton%20HM%5BAuthor%5D&cauthor=true&cauthor_uid=25139174

EB, Cedars M, Rosen M. Comparison of age at natural menopause in BRCA1/2 mutation carriers with a non-clinic-based sample of women in northern California. *Cancer.* 2013 May 1;119(9):1652-9. doi: 10.1002/cncr.27952. Epub 2013 Jan 29. PMID: 23362014; PMCID: PMC3634895.

29. Yang Y, Varvel NH, Lamb BT, Herrup K. Ectopic cell cycle events link human Alzheimer's disease and APP transgenic mouse models. *The Journal of Neuroscience*, 2005; 26 :. 775-784
30. Morel[40] L, Mouse Models of Human Autoimmune Diseases: Essential Tools That Require the Proper Controls, *PLoS Biol.* 2004; 2(8): e241
31. Aizenstein HJ, Nebes RD, Saxton JA, et al. Frequent Amyloid Deposition Without Significant Cognitive Impairment Among the Elderly. *Arch Neurol.* 2008;65(11):1509–1517. doi:10.1001/archneur.65.11.1509
32. 11. Stuart J A.,. Page M M Plasma IGF-1 is negatively correlated with body mass in a comparison of 36 mammalian species *Mechanisms of Ageing and Development* 2010; 131 :591–598

40. http://www.ncbi.nlm.nih.gov/pubmed/?term=Morel%20L%5Bauth%5D

TOWARD TREATING AGING AND EXTENDING LIFE

Treating and preventing aging is the ultimate goal of medicine and the ultimate goal of human emancipation. Any kind of political power a person can achieve is limited by her/his own degradation. In George Orwell's "1984", a political prisoner points to the limit of the immense power a leader in a dictatorship can reach: the decay of his body he can't do anything against. Defeating aging would transform humans into a kind of gods, immortal in the Ancient vision.

And Ancient people were more optimistic in this regard. The epic of Gilgamesh (c. 2150-1400 BCE) depicts the human dream of immortality. The Christian Bible states that "humans will learn to live forever". This message looks very promising. Closer to our times, one of the predictions Nostradamus made for 2015 was that humans will live over 200 years old thanks to the remarkable advances in medicine.

Life-extension- just evolutionarily predictable

But what does tell biology about life extension? A wider view on human evolution may suggest that extending life, treating and preventing aging may be just the next phase in human cultural and biological evolution. Throughout their history, humans have gained more and more autonomy from their environment. This is culture's role; it offers autonomy from resource availability and makes the interaction with the environment less energetically demanding. Culture is a behavioral tool which have reached an incredible level in humans, but it is present in other related (but not only) species. But culture itself is the normal consequence of animal evolution. Because they can move, animals tend toward more autonomy. They chose their environment, to the extent that they can. Culture is just an elongation of animal behavior. And animals actively seek safety, i.e. the maintenance and integrity of their organisms.

Intelligence, the complexity of the nervous system, directly correlates with a longer life expectancy throughout animal evolution. Birds and mammals are considered the animals with the most complex nervous system and behavior. Is there a biological relationship between the evolution of a complex brain and a resilient organism able to confer longevity or does a complex brain lead to a complex behavior that helps preserve the life of the organism for longer?

First of all, we have to differentiate between short and long term organism maintenance. A complex, fine-tuned behavior offers higher chances for survival at every moment. Intelligent animals may fail to die because of the dangerous environment (predators, accidents, starvation) oftener than animals with less complex brains and behaviors. The environment becomes safer and safer as animals become more intelligent. In fact, intelligent species tend to live longer and have fewer offspring. Looking at the animal kingdom, one can notice that there is an inverse relationship between intelligence and fitness, i.e. the capacity of having viable offspring [1]. They live longer, reproduce less and later in life, their development is slower. In this respect, *Homo sapiens* are a champion in a clade (Mammalia), an order (Primata), of species with long and slow lives and reduced fitness.

Like any physical and chemical system, living beings tend to an energetic minimum, in this case, the energetic minimum of the interaction of the organism with the environment. Adaptation is the result of this interaction. Related to the channels of interaction with the environment, various reactions may appear, leading to increasing complexity in some organs and tissue. Epigenetic changes follow the chemical changes in cells, in fact the syntheses in the cytoplasm. Then epigenetic changes lead to mutations and other changes of the genetic material (DNA). Epigenetic changes influence subsequent syntheses for long time, but at the beginning they may be the consequences of metabolic changes. Reaction change, cells change, genes may change. Organisms change. Evolution may be led by energetically favorable reactions, not fitness and spreading genes. And sometimes in chemistry, complex syntheses are energetically favorable. The same principle may underlie the occurrence of complex traits, like intelligence, but not only.

Complexity makes interaction with the environment more specific and energetically cheaper. This vision may explain the evolutionary appearance of complexity, constantly associated, at least in animals, with a decrease in fitness. Is there any kind of genes more frequent in nature, more successful, than bacterial ones? Is there any complex organism fitter than a bacterium? So, following the logic of the selfish gene, complex organisms should not exist. They are the failure of evolution. The pinnacle of failure is the species with the lowest fertility, *Homo sapiens*.

Fitness is a reaction product. If there is no fitness, there is no life, and nothing to talk about. But fitness can't be a life's goal. Chemistry doesn't make plans, chemistry doesn't have goals and strategies. Chemistry works with reactions, their conditions and their products.

The correct question about complex organisms is why these organisms still exist, despite their low fitness. They appeared and still exist because they have a short term survival rate higher than other organisms.

Bacteria are virtually immortal. The signs of aging in bacteria are weak and controversial, related with chemical gradients [2]. Bacteria can live forever, if they don't die because of environmental factors complex organisms hardly feel like stressors. For example, at least some bacteria and other microorganisms are very sensitive to weak ionized radiation (which questions the theory of free radicals in aging). This is a trait disinfection is based upon.

Is there any program for aging?

Although intelligent organisms have long lives, they are not among the species with negligible senescence. Birds and mammals suffer clear degeneration during aging. Organisms with negligible senescence, with no noticeable rise in mortality and decrease in fertility with age [3], are coelenterates (*Hydra*), arthropods or mollusks [4].

The champions of longevity and regeneration among Chordata are not mammals or birds, but fish and reptiles. Why? Is there any genetic program for aging, a new and valuable acquisition of evolution in the complex organism? What is aging in the first place?

For a long time, putative mutations accumulating with age were considered the cause of aging. But there is a problem with such mutations: they were not discovered or they are not as widely spread as proposed [2]. Now, it's certain that there are epigenetic changes with aging, variable in amount in different tissues. But, as mentioned above, epigenetic changes follow chemical changes in the cytoplasm; they follow chemical changes in cells. Life is chemistry, cells and organs consist of chemical reactions. In chemistry, any reaction is virtually reversible. But the equilibrium of these reactions may change. For example, some reaction products are gases or they precipitate and the elimination of a reaction product will alter the equilibrium of the reaction. Aging may result from this alteration of equilibrium. What initiates it? If one thinks about the initial phase of tissue formation, and then after the tissue has been functioning for a long time, it sends and receives different signals, it functions at different rate. Various reactions have different levels. Aging may be the loss of cellular specificity and differentiation. Although homeostasis saves an organisms life for short term, sometimes it can accelerate the chemical alterations responsible for aging. There is no surprise that the champions of longevity or/and regeneration are recruited from poikilothermic species. They simply initiate long chains of reactions with a high degree of conservation. Homeostasis involving hormones may have similar effects, forcing cells to proliferate and differentiate.

According to our biochemical hypothesis of aging [2], this process may result from an organism's functioning. Aging and its rate depends on the capacity of an organism to maintain its specific reactions. Genetics is important in aging only because genes determinate aging in a species, because genes determinate the limits of biochemical functioning.

In chemistry, reactions and their conditions are what matters, not the source of the reactants. So, it is not necessary for a reaction to appear only because it has been taking place for millions years. Longevity may disappear and appear again during evolution. The importance of environmental factors in aging is suggested by the effect of caloric restriction, which increases the lifespan by almost half in various species, from worms and insects to mammals. There is a life span gap between the flying birds and non flying birds with similar body mass. Flying mammals, like bats, have a lifespan similar to flying birds with similar body mass, although in general there is a considerable life span gap between birds and mammals [2].

Toward extending life and immortality in humans

If aging is loss of reaction specificity, stimulating those reactions, restoring their previous level may be the key for preventing, delaying and even curing aging. In this respect, new perspectives on aging are needed, new hypotheses, able to find new biomarkers of aging and treatments based on the predictions of such hypotheses.

References

1. Diaconeasa AG. Civilizația foametei/o alta abordare a

umanizării. Craiova. Sitech. 2007;

2. Daconeasa AG, Spiru L, Turcu I. Human evolution, human brain diseases. Int J Dev Neurosci. 2010; 28: 705.
3. Diaconeasa AG, Rachita M, Stefan-van Staden R-I. A new hypothesis of aging. Med Hypotheses. 2015; 84: 252-257.
4. Finch CE, Austad SN. History and prospects: symposium on organisms with slow aging. Exp Gerontol. 2001; 36: 593-597.
5. Iovita A. The aging gap between species. Create Space Independent

Don't miss out!

Visit the website below and you can sign up to receive emails whenever Amalia Gabriela Diaconeasa publishes a new book. There's no charge and no obligation.

https://books2read.com/r/B-A-YGZGF-OMSBJ

BOOKS 2 READ

Connecting independent readers to independent writers.

Also by Amalia Gabriela Diaconeasa

Civilizația foametei/ O altă abordare a umanizării
Missing Links in Aging
The Civilization of Famine/The human evolution seen from another angle

www.ingramcontent.com/pod-product-compliance
Lightning Source LLC
LaVergne TN
LVHW090414210926
848480LV00002B/560

* 9 7 9 8 2 3 3 9 2 1 9 9 5 *